International Conference on Geological, Geographical, Aerospace and Earth Sciences

(AeroEarth 2013)

IOP Conference Series: Earth and Environmental Science Volume 19

Jakarta, Indonesia
23 – 24 January 2013

Editors:

F. Gaol
B. Soewito
A. Desai

ISBN: 978-1-63439-960-9
ISSN: 1755-1307

Printed from e-media with permission by:

Curran Associates, Inc.
57 Morehouse Lane
Red Hook, NY 12571

Some format issues inherent in the e-media version may also appear in this print version.

Printed by Curran Associates, Inc. (2015)

For permission requests, please contact the Institute of Physics
at the address below.

Institute of Physics
Dirac House, Temple Back
Bristol BS1 6BE UK

Phone: 44 1 17 929 7481
Fax: 44 1 17 920 0979

techtracking@iop.org

Additional copies of this publication are available from:

Curran Associates, Inc.
57 Morehouse Lane
Red Hook, NY 12571 USA
Phone: 845-758-0400
Fax: 845-758-2634
Email: curran@proceedings.com
Web: www.proceedings.com

TABLE OF CONTENTS

IOP Conf. Series: Earth and Environmental Science **19** (2014) 011001 doi:10.1088/1755-1315/19/1/011001

Preface

The 2013 International Conferences on Geological, Geographical, Aerospace and Earth Sciences (AeroEarth 2013), was held at the Swiss Bell Mangga Besar, Jakarta, Indonesia, on 23 December 2013.

The AeroEarth conference aims to bring together researchers, engineers and scientists in the domain of interest from around the world. AeroEarth 2013 promotes interaction between the theoretical, experimental, and applied communities, so that high-level exchange is achieved in new and emerging areas within Earth Science. Through research and development, earth scientists have the power to preserve the planet's different resource domains by providing expert opinion and information about the forces which make life possible on Earth.

We would like to express our sincere gratitude to all in the Technical Program Committee who have reviewed the papers and developed a very interesting Conference Program as well as the invited and plenary speakers.

This year, we received 91 papers and after rigorous review, 17 papers were accepted.

The participants come from 8 countries. There are 3 (three) Plenary Sessions and two invited Speakers.

It is an honour to present this volume of IOP Conference Series: Earth and Environmental Science (EES) and we deeply thank the authors for their enthusiastic and high-grade contribution.

Finally, we would like to thank the conference chairmen, the members of the steering committee, the organizing committee, the organizing secretariat and the financial support from the conference sponsors that allowed the success of AeroEarth 2013.

The AeroEarth 2013 Proceedings Editors

Dr. Ford Lumban Gaol
Dr. Benfano Soewito
Dr. Amit Desai

IOP Conf. Series: Earth and Environmental Science **19** (2014) 011001 doi:10.1088/1755-1315/19/1/011001

Keynote title:
Modeling Animal Waste Pathogen Indicators Transport from Agriculture land to Streams

Speaker:
Pramod Pandey, Population Health and Reproduction Department, School of Veterinary Medicine, University of California, Davis, California 95616, USA

Synopsis:
The transport of animal waste pathogens from crop land to streams potentially elevates pathogen levels in stream water. Animal waste is commonly applied into crop land as fertilizers to enhance the nutrient levels of the crop land; however, manure application into the crop land causes pathogen problems in stream water. Elevated pathogen levels in ambient water result potential health risk to human. In order to control pathogen levels in ambient water bodies, improving understanding of pathogen transport from the crop land to water bodies at watershed scale is required. Here we investigate on pathogen indicators transport from the clop land receiving manure as fertilizers to streams. A watershed scale hydrological model was applied to estimate the transport of pathogens from crop land to streams. Extensive pathogen indicator monitoring at watershed scale was carried out to verify the model predictions. With certain assumptions, model results are reasonable. This study provides guidelines for model development capable of predicting pathogen/pathogen indicators to ambient water bodies. The proposed study will be helpful understanding the impacts of animal waste on ambient water quality.

Keywords: watershed, hydrological model, pathogen indicator, animal waste, crop land

About the Speaker: Pramod Pandey is working as a specialist in Population Health and Reproduction Department at University of California, Davis, USA. Currently his research focus is controlling microbial pollution of animal waste and mitigating public health risk. He has done extensive field scale and modeling research on understanding stream water quality, animal waste treatment, watershed hydrology, and renewable energy production from animal waste. He has developed models for improving water resources in rural communities of humid and arid regions. Pramod holds PhD in Agricultural Engineering from Department of Agricultural and Biosystems Engineering (ABE), Iowa State University, Ames, Iowa, USA. Dr Pramod has Graduate Certificate in Geographic Information Systems (GIS) from Department of Community and Regional Planning (CRP), Iowa State University, Ames, Iowa, USA and Master of Science (M.S.) in Agricultural Engineering from Department of Biological Systems Engineering (BSysE), Washington State University, Pullman, Washington, USA as well as Master of Technology (M.Tech.) in Aquaculture Engineering from Department of Agricultural and Food Engineering (AGFE), Indian Institute of Technology, Kharagpur, India.

IOP Conf. Series: Earth and Environmental Science **19** (2014) 011001 doi:10.1088/1755-1315/19/1/011001

Title :
High Throughput satellites and its future roles on Earth

Speaker :
Dr. Ing. Arifin Nugroh, PT Citra Sari Makmur

About the Speaker: Dr Arifin Nugroho is one of the pioners in Satellite Industry in Indonesia. He was working with some Governent projects in Satellites that mainly in Remote Sensing and Telecommunications. He Holds Dr. Ing from Universite de Rennes I, France. He was retired from PT Telkom Indonesia, the biggest Telecommunication company in Indonesia after work there for more than 31 years.

Workshop

Title: Remote Sensing and its Implementation using GPS Tool Kits
Speaker: Dr. Soetanto Suroto
Affiliation: Lembaga Telematika Indonesia
Abstract: In this workshop, the participants will be taught about the incorporated and applied concepts of GPS tool kits.

IOP Conf. Series: Earth and Environmental Science **19** (2014) 011001 doi:10.1088/1755-1315/19/1/011001

Crop land as a source of pathogens

1 Why is the
polycentrism a desirable
urban stucture?

IOP Conf. Series: Earth and Environmental Science **19** (2014) 011001 doi:10.1088/1755-1315/19/1/011001

The Swiss Bell Mangga Besar Hotel – Lavender II room at 3rd floor Jakarta, 23 December 2013

Peer review statement

All papers published in this volume of *IOP Conference Series: Earth and Environmental Science* have been peer reviewed through processes administered by the proceedings Editors. Reviews were conducted by expert referees to the professional and scientific standards expected of a proceedings journal published by IOP Publishing.

Comparison Between Roe's Scheme and Cell-centered scheme For Transonic Flow Pass Through a Turbine Blades Section

Abobaker Mohammed Alakashi[1, a]**,Bambang Basuno**[2, b]
[1, 2] Faculty of Mechanical and Manufacturing Engineering, University Tun Hussein Onn Malaysia (UTHM), Parit Raja, Batu Pahat, 86400 Johor, Malaysia

[a] Abobaker.alakashi@gmail.com, [b]Bambangb@uthm.edu.my

Abstract. The finite volume method (FVM) is a discretization technique for partial differential equations, especially those that arise from physical conservation laws. FVM uses a volume integral formulation of the problem with a finite partitioning set of volumes to discretize the equations[1]. The present work employs the Cell- centred and Roe's scheme to investigate the inviscid compressible flow pass through a two dimensional blade on two types of blade shapes: Blazek and SE1050 blade models. The governing equation of fluid motion of the flow problem in hand is the Euler Equation. The behavior of this equation will depend on the local Mach number if the governing equation stated in the form as the governing equation of steady flow problem. If the local Mach number is less than one, the governing equation will behave as elliptic type of differential equation while if the Mach number is greater than one it will behave as hyperbolic type of differential equation. Elimination of the presence of those two types governing equation for the case of transonic flow problem can be achieved by representing the Euler equation in unsteady form. So the equation becomes hyperbolic with respect to time. There are various Finite Volume Methods that can used for solving hyperbolic type of equation, such as Cell-centered scheme [2], Roe Upwind Scheme [3] and TVD Scheme[1]. Those computer codes apply in the case of inviscid two dimensional compressible flow pass through blades of turbine. The present work focuses on two computer codes, first based on Cell Centre scheme and the second one based on Roe's scheme.

1. Introduction

In the absence of viscous effects, the governing equation of fluid motion pass through an arbitrary body can be represented by Euler equations. For two dimensional compressible flow problems, this governing equation of motion in the form of non linear differential system equation. As result no analytic solution can be found for such kind of flow problems and a numerical approach is required. The governing equation of fluid motion can be presented in integral form which allow one to apply spatial descretization through the use of finite volume approach. There are various Finite Volume Methods can used for solving hyperbolic type of equation, such as Cell-centered scheme [2], Roe Upwind Scheme [3] and TVD Scheme[1]. Basically those three type finite volume methods are able to resolve some of the difficulties are often faced by the other methods which may required a fine and structured grid such as in the finite difference based method [2].

As the method of solving the flow problem through the governing equation of fluid motion in integral form make the finite volume method has a large number of options for the defining the control volumes around which the conservation laws are expressed. Modifying the shape and location of the control volumes associated with a given mesh point, as well as varying the rules and accuracy for the evaluation of the fluxes through the control surfaces, gives considerable flexibility to the finite volume method[4]. In addition to this, through directly discretizing to the integral form of the conservation laws, one can ensure that the basic quantities mass, momentum and energy will also remain conserved at the discrete level [3]. The main advantage of the finite volume method is that the spatial discretisation is carried out directly in the physical space. Thus, there are no problems with any transformation between coordinate systems [1,2].

2. Governing Equation of fluid motion in The integral conservation laws

As in usually adopted in flow analysis , let the flow variables are denoted by p, ρ, u, v, E and H which they are representing the pressure, density, Cartesian velocity components, total energy and total enthalpy respectively. For a perfect gas one has .

$$E = \frac{p}{(\gamma-1)\rho} + \frac{1}{2}(u^2 + v^2) \quad \text{and} \quad H = E + \frac{p}{\rho} \tag{1}$$

Where γ is the ratio of specific heat. The Euler equations for two dimensional inviscid flows can be written in integral form for a region Ω with boundary $d\Omega$ as:

$$\frac{\partial}{\partial t}\iint_{\Omega} \vec{U} dxdy + \oint_{d\Omega} \left(\vec{F}dy - \vec{G}dx\right) = 0 \tag{2}$$

Where x and y are Cartesian coordinates and

$$\vec{U} = \begin{pmatrix} \rho \\ \rho u \\ \rho v \\ \rho E \end{pmatrix}, \quad \vec{F} = \begin{pmatrix} \rho \\ \rho u^2 \\ \rho uv \\ \rho uH \end{pmatrix}, \quad \vec{G} \begin{pmatrix} \rho \\ \rho uv \\ \rho v^2 + p \\ \rho vH \end{pmatrix}$$

The discretization procedure follows the method of lines in decoupling the approximation of the spatial and temporal terms. The computational domain is divided into quadrilateral cells as in the sketch, and a system of ordinary differential equations is obtained by applying equation 2 to each cell separately. The resulting equations can then be solved by several alternative time stepping schemes. In general if applied the conservation laws on the Cell ABCD as figure (1)

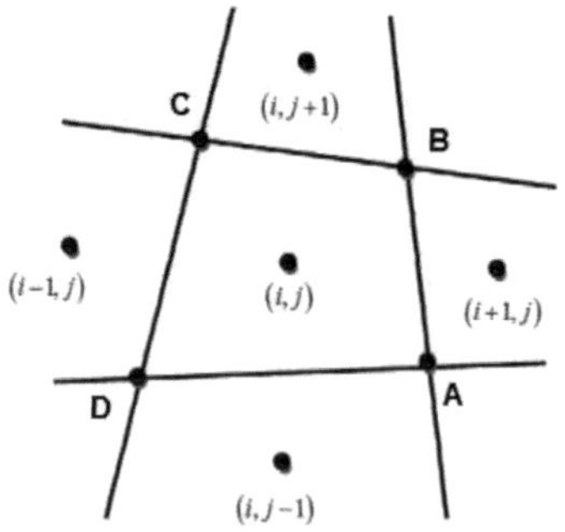

Figure 1. Cells in the finite-volume approach

One need to define the parameter geometry of the grid cell. The grid cell ABCD as shown in above figure has a side AB surface vector $\bar{S}_{AB}$ and its normal direction $\hat{j}_{AB}$ defined as :

$$\bar{S}_{AB} = \Delta y_{AB}\hat{i} - \Delta x_{AB}\hat{j} \quad \text{and} \quad \hat{j}_{AB} = (y_B - y_A)_i - (X_B - X_A)_j$$

While the area of grid cell ABCD [1,2]. Can be obtained as:

$$\Omega_{ABCD} = \frac{1}{2}\left|\vec{X}_{AC} - \vec{X}_{BD}\right| \;; \qquad \text{Where} \quad \vec{X}_{AB} = \vec{X}_B - \vec{X}_A$$

Or in term of coordinate point (X,Y) as :

$$\Omega_{ABCD} = \frac{1}{2}\left[(X_C - X_A)(y_D - y_B) - (X_D - X_B)(y_C - y_A)\right] \tag{3}$$

In manner to determine the flux vector can be done in various approaches. The flux F crossing the side surface AB denoted as $\vec{F}_{AB}$ can be done with one of following approach.

- Average of fluxes $F_{AB} = F\left(\frac{U_{ij} - U_{i+1,j}}{2}\right)$
- Flux of the average flow variable $F_{AB} = F\left(\frac{U_{ij} + U_{i+1,j}}{2}\right)$
- Average of fluxes in A&B $F_{AB} = \frac{1}{2}(F_A + F_B)$

In the last approach, approach c, the flux F_A determine firstly by defining that the flow variable U at A determine as :

$$U_A = \frac{1}{4}\left(U_{ij} + U_{i+1,j} + U_{i+1,j-1} + U_{i,j+1}\right) \tag{4.a}$$

Then the average the fluxes F_A becomes

$$F_A = \frac{1}{4}\left(F_{ij} + F_{i+1,j} + F_{i+1,j-1} + F_{i,j+1}\right) \tag{4.b}$$

With the ingredient in manner how to define the flux vector F as well as G as above, the numerical approach for solving the flow problem by using a finite volume method becomes solving the problem of ordinary differential equation with respect to time in the form of equation as given below;

$$\frac{\partial}{\partial t}U_{ij}\Delta X\Delta y + \left(F_{i+\frac{1}{2},j} - F_{i-\frac{1}{2}}\right)\Delta y + \left(g_{i,j+\frac{1}{2}} - g_{i,j\frac{1}{2}}\right)\Delta X = Q_{ij}\Delta X\Delta y \tag{5}$$

In part of the Roe's Scheme approach, The finite-volume form of the Euler equations can be written as;

$$\frac{dU}{dt} = \frac{1}{A}\Sigma_{faces}(F\Delta y - G\Delta x) \tag{6}$$

Where A is the area of that cell Δx and Δy are the changes of x and y along face

Defining the face length and normal and tangential velocities as;

$$\Delta s = \sqrt{(\Delta x)^2 + (\Delta y)^2}, \quad u_n = \frac{(u\Delta y - v\Delta x)}{\Delta s} \quad \text{and} \quad u_t = \frac{(u\Delta x - v\Delta y)}{\Delta s}$$

The flux through a face may be written as:

$$(F\Delta y - G\Delta x) = \begin{pmatrix} \rho u_n \\ \rho u_n u + p\frac{\Delta y}{\Delta s} \\ \rho u_n v - p\frac{\Delta x}{\Delta s} \\ \rho u_n H \end{pmatrix} \Delta s \cong \emptyset\,\Delta s \tag{7}$$

The flux through a face is a function of the values at the face midpoint, given by the reconstruction in the cells to the " left" and "right" of the face. Using Roe's approximate Riemann solver, this flux function is;

$$\emptyset(U_L, U_R) = \frac{1}{2}[\emptyset(U_L) + \emptyset(U_R)] - \frac{1}{2}\sum_{k=1}^{4} |\hat{a}_k|^* \Delta V_k \hat{R}_k \tag{8}$$

$$\hat{a} = \begin{pmatrix} \hat{u}_n - \hat{c} \\ \hat{u}_n \\ \hat{u}_n \\ \hat{u}_n + \hat{c} \end{pmatrix} \quad ; \quad \Delta V = \begin{pmatrix} \frac{\Delta p - \hat{\rho}\hat{c}\,\Delta u_n}{2\hat{c}^2} \\ \frac{\hat{\rho}\Delta u_t}{\hat{c}} \\ \Delta\rho - \frac{\Delta p}{\hat{c}} \\ \frac{\Delta p - \hat{\rho}\hat{c}\,\Delta u_n}{2\hat{c}^2} \end{pmatrix} \quad \text{and}$$

$$\hat{R} = \begin{pmatrix} 1 & 0 & 1 & 1 \\ \hat{u} - \hat{c}\frac{\Delta y}{\Delta s} & \hat{c}\frac{\Delta x}{\Delta s} & \hat{u} & \hat{u} + \hat{c}\frac{\Delta y}{\Delta s} \\ \hat{v} + \hat{c}\frac{\Delta x}{\Delta s} & \hat{c}\frac{\Delta y}{\Delta s} & \hat{v} & \hat{v} - \hat{c}\frac{\Delta x}{\Delta s} \\ H - \hat{u}_N\hat{c} & \hat{u}_t\hat{c} & \frac{\hat{u}^2 + \hat{v}^2}{2} & H + \hat{u}_n\hat{c} \end{pmatrix}$$

Where;

$$\hat{\rho} = \sqrt{\rho_L \rho_R} \quad \hat{u} = \frac{\sqrt{\rho_L} u_L + \sqrt{\rho_R} u_R}{\sqrt{\rho_L} + \sqrt{\rho_R}} \quad \hat{v} = \frac{\sqrt{\rho_L} v_L + \sqrt{\rho_R} v_R}{\sqrt{\rho_L} + \sqrt{\rho_R}} \quad \hat{H} = \frac{\sqrt{\rho_L} H_L + \sqrt{\rho_R} H_R}{\sqrt{\rho_L} + \sqrt{\rho_R}}$$

Where $\hat{c}, \hat{u}_n,$ and $\hat{u}_t$ are calculated directly from $\hat{\rho}, \hat{u}, \hat{v},$ and $\ddot{H}$. The flux difference terms (the summation in Eq. (38)) provide the upwind character which stabilizes the scheme. To prevent expansion shocks, an entropy fix imposed [2.5]. A smoothed value, $|\hat{a}^{(k)}|^*$, is defined to replace $|\hat{a}^{(k)}|$

For the two acoustic waves (k = 1, k = 4). For those two waves,

$$|\hat{a}^{(k)}|^* = \begin{cases} |\hat{a}^{(k)}|, & |\hat{a}^{(k)}| \geq \frac{1}{2}\delta a^{(k)} \\ \frac{(\hat{a}^{(k)})^2}{\delta a^{(k)}} + \frac{1}{4}\delta a^{(k)}, & |\hat{a}^{(k)}| \leq \frac{1}{2}\,\delta a^{(k)} \end{cases} \tag{9}$$

$$\delta a^{(k)} = max\left(4\Delta a^{(k)}, 0\right), \qquad \Delta a^{(k)} = a_R^{(k)} - a_L^{(k)}$$

For each cell, the face fluxes, calculated as above, are summed to give the residual for the cell [2.6],

$$\text{Res(U)} = -\frac{1}{A}\sum_{\text{faces}} \emptyset \Delta s \tag{10}$$

3. Result and Discussion

The present work employs the Cell- centred and Roe's scheme to investigate the inviscid compressible flow pass through two dimensional blade on two type of blade shapes Blazek and SE1050 blades. The grid cells of the cases follow structured grid system The grid cells of the problem in hand as shown in Figure 2 and 3. Two computer codes of solver had been developed, the first one created base on Cell- centred scheme by using a structured grid while the second one based on Roe's scheme by using a structured grid approach also . In the incoming flow speed is around 300 m / sec at standard sea level, the flow problem can be considered as problem of the transonic flow around a turbine blade cascade. The boundary condition applied to this flow problem is given as follows Stagnation pressure : 100000 Pa ,Static pressure 99448.5 Pa, The velocity component in x direction 300.4 m/sec, The velocity component in y direction 0.0 m/sec, Static temperature T = 300 K, The exit pressure Pe = 52830 Pa.

Figure 4,a,b..d. below describe the Comparison of Mach number distribution in domain of interest of cross section Blazek and SE1050 turbine blades by using computer code of Cell- centred and Roe's Schemes on lower and upper parts respectively.

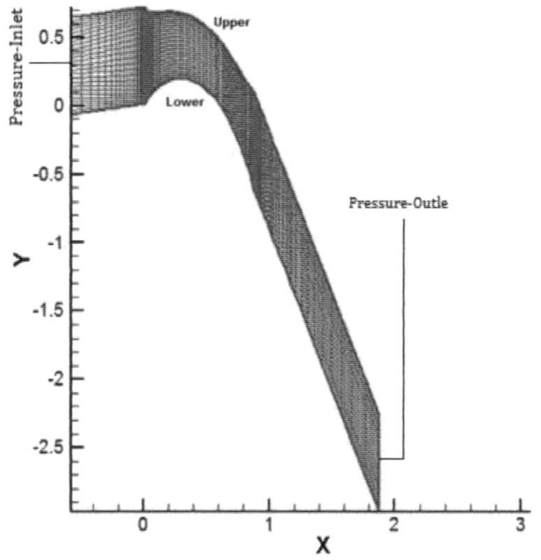

Figure 2. Cross section of Blazek turbine blades

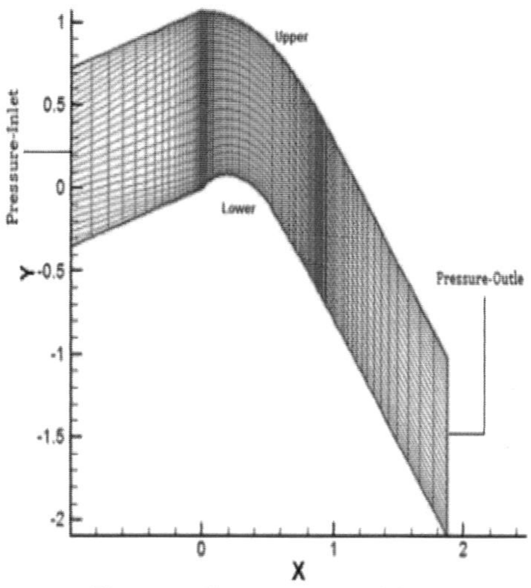

Figure 3. Cross section of SE1050 turbine blades

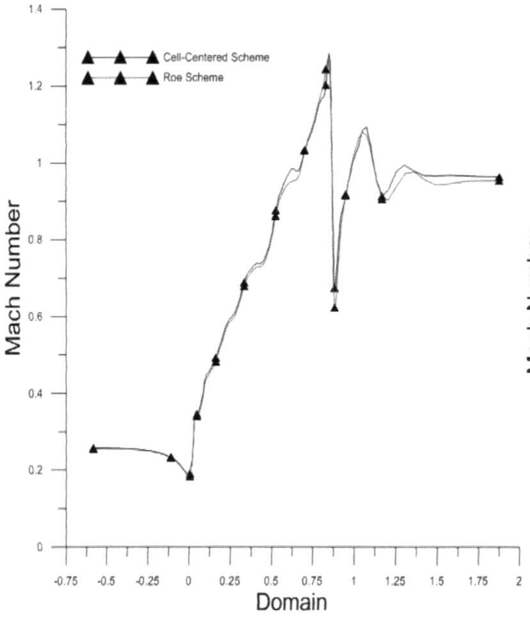

Figure 4.a. Mach number at lower part for Blazek blade.

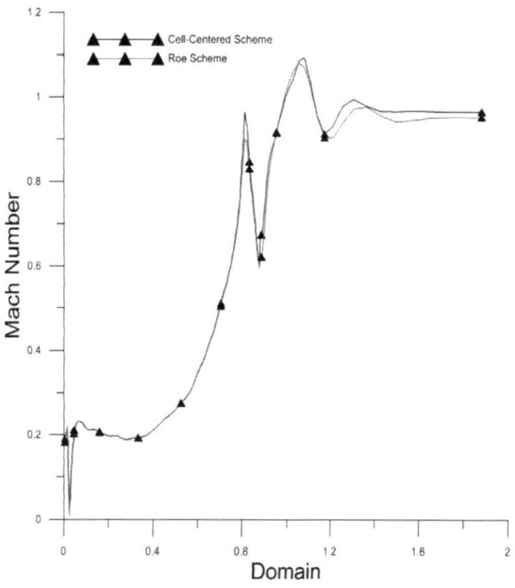

Figure 4.b. Mach number at upper part for Blazek blade.

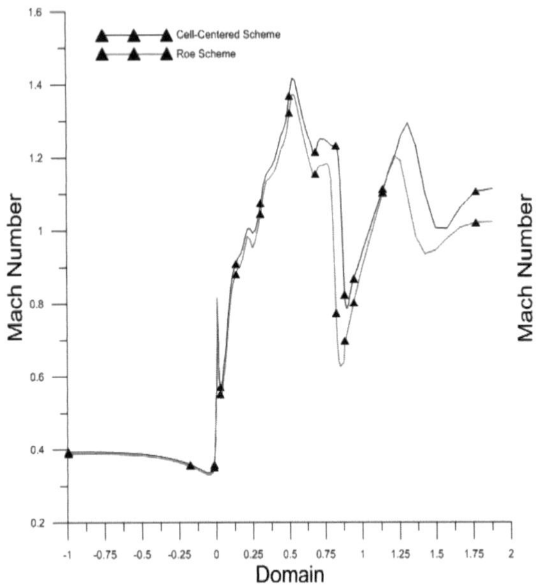

Figure 4.c. Mach number at lower part for SE1050 blade.

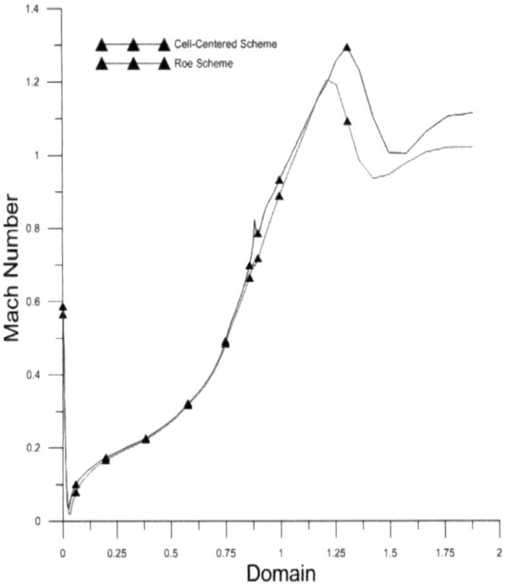

Figure 4.d. Mach number at Upper part for SE1050 blade.

Conclusion:

From the graph as shown above, the results in term of Mach number distribution over blade surfaces (upper and lower part) , for the case of Blazek blade, both numerical schemes give almost the same result. While for the case of SE1050 blade, a slightly different result occurred in the region near and after the blade trailing edge. However over the whole blade surface both numerical scheme provide their results in a good agreement each to other. From this point, it can be concluded the present Cell-centred Scheme and Roe's Scheme may represents the numerical scheme which can be extended in the use for further study on fluid behaviour over a complete turbine system which consist of "stator and rotor" sections. This combined system is on going in the present research.

References:

[1] K.. A . Hoffman and S.T .Chiang, Computational-Fluid-Dynamics Volume I, Engineering Education System.," USA, 2000.

[2] J. Blazek, Computational Fluid Dynamics Principles and Applications, 2nd Ed.," Switzerland, 2001.

[3] J. H. Ferziger, and M. Perić, Computational Methods for Fluid Dynamics.," Springer, 1996

[4] J. D. Anderson, Jr, Computational Fluid Dynamics, the basics with applications, McGraw-Hill, 1995.

[5] V. Hannemann, and others, Dissipation of upwind schemes at high wave numbers.," DLR - Institute of Aerodynamics and Flow Technology, Germany, 2012.

[6] G. Herve and Angelo Murrone, On the behavior of upwind schemes in the low Mach number limit: II. Godunov type schemes.," France, 2003

2013 International Conferences on Geological, Geographical, Aerospace and Earth Sciences IOP Publishing
IOP Conf. Series: Earth and Environmental Science **19** (2014) 012002 doi:10.1088/1755-1315/19/1/012002

Constraining Stochastic Inversion with Frequency Domain Seismic Signature for Seismically Thin-bed Interpretation

EkoWidi Purnomo and Deva Prasad Ghosh

Center for Seismic Imaging, Faculty of Petroleum Geosciences UniversitiTeknologi PETRONAS Malaysia

ekowidip@gmail.com

Abstract. An alternative technique in interpreting thin-bed structure has been developed. The technique involved stochastic inversions which use frequency domain energy spectral attribute as a constraint instead of time domain seismic amplitude. Maximum Amplitude Weighed Integrated Energy Spectra is a proposed energy spectral attribute which was used to constrain the stochastic process. Amplitude Weighed Integrated Energy Spectra is a deployed seismic attribute obtained by multiplying integrated energy spectra with maximum amplitude of a seismic trace. It is shown that Amplitude Weighed Integrated Energy Spectra provides a more separable signature in responding to bed thickness changes than seismic signature. A lower degree of ambiguity of Amplitude Weighed Integrated Energy Spectra in sensing thin-bed seismic is a potential method of reducing thin bed interpretation uncertainty.Qualitatively, Amplitude Weighed Integrated Energy Spectra is capable of showing one of the reported very thin meandered channel complexes of gas reservoir of Stratton field which is difficult to be seen in seismic amplitude. In this research, Amplitude Weighed Integrated Energy Spectra is incorporated in a stochastic seismic inversion to improve both accuracy and precision (certainty) of thin-bed interpretation. The signature of Amplitude Weighed Integrated Energy Spectra is used to constrain the degree of match (likelihood) between seismic model and data. Synthetic data testing shows that the proposed method significantly improves both accuracy and precision of a single wedge model seismic inversion. The thickness and reflection coefficient are estimated more accurately, although limited information is used. The proposed method was tested to invert a structurally subtle gas production zone of Stratton field. Confirmed by well log data, a cross section of inverted impedance showed that some channel complex structure of gas reservoirs are able to be imaged.

1. Introduction

Reservoir characterization is subject to uncertainty. One of the biggest uncertainty contributors is structural interpretation. Seismic inversion is a widely used tool to define reservoir structure. However, the band limited frequency character of seismic data precludes seismic to recover all frequency band of reservoir. Interpretation of reservoir structure beyond the seismic band is uncertain and difficult. Such structure can not be resolved by seismic. Thin bed is defined as reservoir with thickness below the seismic resolution. When the reservoir thickness goes to below seismic resolution (tuning), the seismic response are remain unchanged. For any reservoir thinner than the tuning thickness, there are infinite numbers of reservoir thickness and impedance combinations that give the same seismic amplitude.

Stochastic inversion [1]is a model-based seismic inversion which commonly used to handle the reservoir characterization uncertainty. Stochastic inversion uses a forward model to generate synthetic seismic data as part of the inversion algorithm. Several or may be a lot of different synthetic seismic are generated and then comparing to the real seismic to be inverted. The 'good fit' synthetic seismic, then are kept as output of the inversion and the 'bad fit' one are discarded. There are different families of stochastic inversion algorithms, from simple acceptation/rejection schemes to more complex Bayesian formulation [2].

To model a higher-than-seismic frequencies, where thin-bed structure associate with, a blocky model earth's impedance is described. This is not valid assumption actually; however uniqueness issue is come from here. Theoretically, stochastic inversion outputs all of the acceptable property models (or a representative distribution of those models) [1]. But, reported by Gunning et.al [2], that stochastic inversion still has problem when inverting thin-bed structure. Thin-bed structures are precluded in stochastic seismic inversion result.

This paper aims to improve the uniqueness (certainty) of thin-bed interpretation and its accuracy as well. The aim will be accomplished through three step of work i.e. deploying energy spectral-based thin-bed seismic attribute, developing thickness-reflectivity coefficient modeland constructing a stochastic inversion with spectral attribute used as constraint.

2. Theory and Methodology

2.1. Maximum Amplitude Weighted Integrated Energy Spectra Deployment

The bed thickness is related to energy spectra of seismic. As the peak frequency is shifted when the bed going below seismic tuning, the integrated energy spectra (INTENS) can consistently detect the bed thickness [3]. Mathematically, an Integrated Energy Spectra of a time series data at frequency f is defined as:

$$E(f) = 100\,\frac{\int_{fl}^{f} A(f)A^{*}(f)df}{\int_{fl}^{fu} A(f)A^{*}(f)df} \qquad (1)$$

where A(f) and A*(f) are the amplitude spectra of the time series at frequency f and the conjugate consecutively. f_l and f_u is the lowest and highest frequency value consecutively.

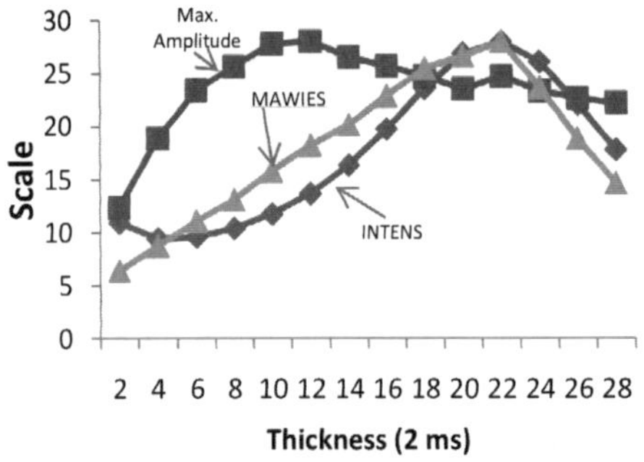

Figure 1. Bed-thickness response of Maximum Amplitude response (rectangled curve), Integrated Energy Spectra (diamoned curve) and Maximum Amplitude Weighed Integrated Energy Spectra (triangled curve).

Maximum weighed integrated energy spectra (MAWIES) is a new seismic attribute which is developed based on previous technique of estimating thin-bed thickness [4][5] and seismic attribute combination [6][7]. The attribute is product of integrated energy spectra (Eq. 1) and its corresponding maximum amplitude. Mathematically, the attribute can be expressed as:

$$E_m(f) = A_m \times E(f) \tag{2}$$

whereE_m is the new attribute, A_m is maximum amplitude, E is Integrated Energy Spectra and f is frequency.

Comparing with the integrated energy spectra, the new attribute has more linear response to layer thickness changes (Fig 1). It indicates that Maximum Amplitude Weighed Integrated Energy Spectra is a good probe of bed thickness.

2.2. Constraining Stochastic Inversion with Maximum Amplitude Weighed Integrated Energy Spectra

The stochastic inversion proposed is build base on Bayesian theorem as following

$$\mathbf{P(RC, d)|E_m)} = \frac{\mathbf{P(E_m|RC,d)\ P(RC,d)}}{\mathbf{P(E_m\ )}} \tag{3}$$

Where P() expresses the probability density function pdf. RC refers to Reflectivity Coefficient between layer interfaces, d refer to thickness of layer. P(RC,d|E_m) is posterior model of RC and d, as the result of the inversion. P(E_m|RC,d) is likelihood model which defines degree of fit between Maximum Amplitude Weighed Integrated Energy Spectra of true seismic with Maximum Amplitude Weighed Integrated Energy Spectra of synthetic seismic. The synthetic seismic is generated by convolution between reflectivity model (combination of RC and thickness) with an assigned wavelet. P(RC,d) is prior model of modeled reflectivity. And, P(E_m) is distribution function of true Maximum Amplitude Weighed Integrated Energy Spectra. The inversion aim to solve the Eq. 3 including analysis of posterior distribution resulted. Markov Chain Monte Carlo is used to populate the posterior model.

3. Result and Discussion

3.1. Single Trace Synthetic Model Inversion

The proposed method has tested to invert a seismic trace from a thin porous sand with velocity of 3048 m/s and density of 2.3 g/cc encased in a non porous sand with velocity of 4267 m/s and of 2.502 g/cc. The time thickness of encased is 4 ms, while the seismic tuning thickness [4] is about 16 ms.Figure 7.(i) shows 'spaghetti' plot of the inversion.

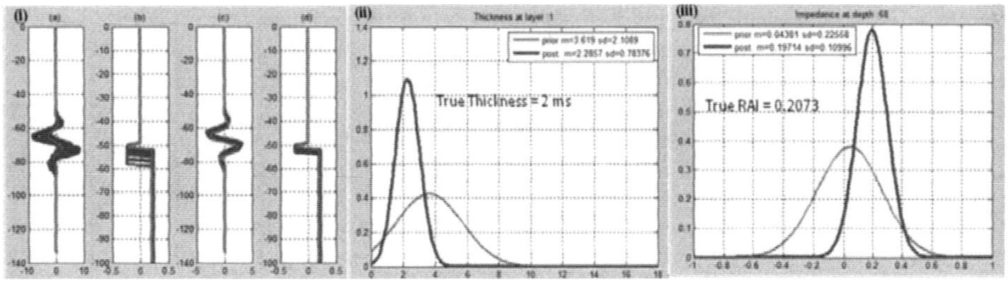

Figure 2. 'Spaghetti plot' of stochastic inversion (i). (a) Real seismic (red) between Prior model (blue); (b) true relative acoustic impedance (red) among20 running sum of modeled relative impedance; (c)real relative acoustic impedance (red) among20 modeled relative impedance; (d) real realtive impedance (red) overlaying20 inverted relative impedance (blue); (ii) Distribution plot of thickness and (iii)Distribution plot of relative impedance at time level of 66 ms (encased porous sand) Figure 2 (i).

Comparing to the prior model (before inversion) Figure 2.(i).(a) and (b), the narrower blue curves in spaghetti of posterior model (after inversion) Figure 2.(i). (c) and (d), indicate that the acuracy and precision of thickness and impedance are increased during the inversion.

Figure 2.(ii) is a quantitative analysis of posteriors model resulted in Figure 2.(i). The analysis is commonly called marginalization. Twenty thickness estimations are extracted from both of twenty posterior and twenty prior models. The distributions of estimated thickness are then plotted and analyzed. Distribution's mean of 3.0952 and standard deviation of 1.8683 is resulted from prior model. In different, mean of 2.8571 and standard deviation of 0.6547 are resulted from posterior distribution.The mean of distribution expresses the thickness estimation and the standard deviation quantifies the uncertainty or the estimation error. Comparing to the true thickness of 2, the mean of posterior distribution is closer to the true thickness than the prior's. It indicates that the inversion increases the accuracy of thickness estimation. The smaller standard deviation of posterior distribution than prior's informs that the inversion improve the precision in other word reduce the uncertainty. Figure 2.(iii)shows analyzes of impedance estimation at time depth level of encased porous sand. From a real impedance of -0.2073, prior model estimates -0.04381 with standard deviation 0.22558 and posterior model estimates -0.19714 with standard deviation of 0.10996. Again, the distribution shows that the accuracy and precision of relative impedance estimation are increased by the inversion.

3.2. Real Seismic Data Inversion

The proposed stochastic inversion method has been tested to invert the gas production zone 3D seismic of Stratton field [8][9]. The inverted seismic covers a time window from 1633 to 1700 millisecond time depth, where which two very close F37 and F39 reservoir are located. Hardage et.al [8] reported that the F37 reservoir was approximately 20 ft (6 m) above the F39, which is only 4 ms (2 seismic sampling rates) two-way travel time different. The shallowest and deepest marker of F37 reservoir is identified 1633 and 1667 millisecond, respectively. While the F39 reservoir is at depth 1639 and 1673 millisecond. Through a static pressure study, Hardage et.al [8] found that both two reservoirs are segregated into at least 4 compartments. All compartments are very thin and difficult to be seen in seismic data.Twenty realizations have been inference from every twenty prior impedance sample. Figure 3 shows a cross section of one of realizations cube. The cross section is drawn along cross line number 154 where the well number 7 (Well#7) and 8 (Well#8) located. Well#7 marks the F37 reservoir at 1631 ms and 1639 ms for F39 reservoir, while Well#8 marks the F37 reservoir at 1644 ms and 1653 ms for F39 reservoir. It is shown very interesting in Figure 3 (shown by white ellipses), that both the well markers are coinciding with relatively similar inverted impedance value.

Another interesting figure is shown by Well#10. Well#10 is located at cross cline number 158, 4 trace intervals different with Well#7 and Well#8 locations. It is interesting because the F37 and F39 marker at Well#10 (1650 and 1662 ms) coincide with similar impedance value at Well#7 and Well#8. This leads a lateral extension interpretation of F37 and F39 reservoir from cross line number 154 to 158 along in line number of Well#10. It can be concluded that the F37 and F39 reservoir are identified by this inversion and possible to further characterization.

Figure 4 shows the probability analysis of 'net reservoir volume' estimation of F37 and F39 reservoirs when 0 to 0.05 relative AI range is taken as reservoir character. That is very rough volume estimation. The estimation calculates the run sum impedance ratio between reservoir and non reservoir from whole volume of inverted impedance. The ratios are calculated from all of twenty realization volume and the distribution then analyzed. The estimated P90 'net-volume' is the "pessimistic" 'net-volume' that is exceeded by 90% of the realizations (there is a 90% chance the 'net-volume' will be achieved). The estimated P10 reservoir 'net-volume' is the "optimistic" 'net-volume' that is exceeded by only 10% of the realizations. There might be many very different realizations that give that P90 or that P10 'net-volume'

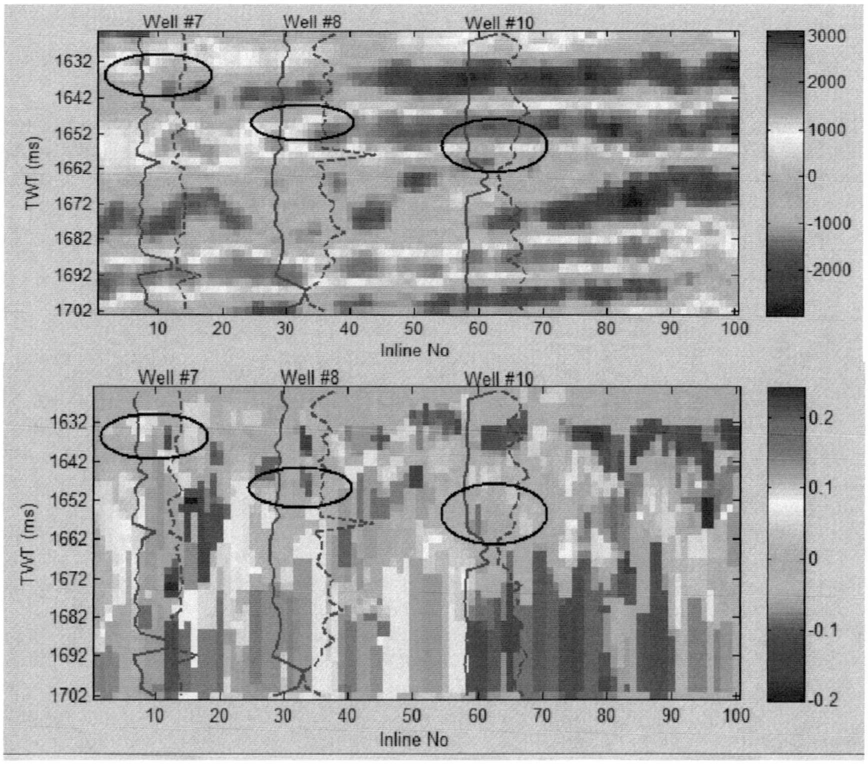

Figure 3. Cross section of cross-line 154 of inverted F37-F39 reservoir seismic. Seismic section (above) and Inverted Relatif AI (below). Wellog data are Spontaneous Potential (SP), full-line, and Gamma Ray (GR), dash-line.Black Ellipses indicate reservoir's well marker.

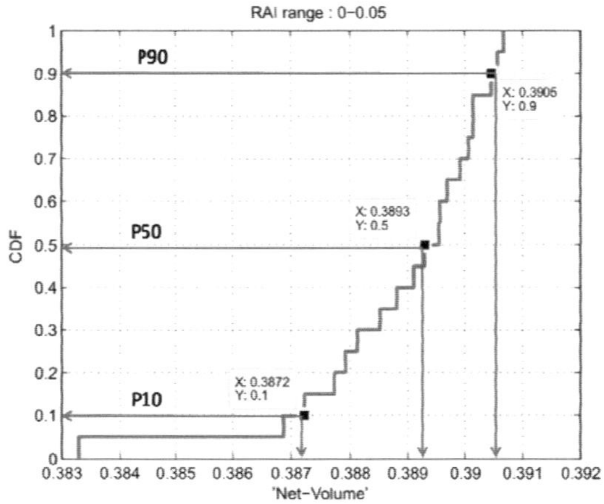

Figure 4. Cumulative Density Function of reservoir's 'Net-Volume' if range of 0 to 0.05 runsum AI is taken for reservoir character.

4. Conclusion

A new energy spectral attribute has been deployed. The deployment is done by combining the rapid changes of maximum absolute amplitude and wide range linearity of integrated energy spectra in responding thin-bed thickness changes. The new attribute is a product of integrated energy spectral and its maximum absolute amplitude. The attribute is called Maximum Amplitude Weighted Intrgrated Energy Spectra (Maximum Amplitude Weighed Integrated Energy Spectra). The multiplication makes integrated energy spectra weighted and, in turn, makes its signature more separable to detect thin-bed thickness changes. The derivative of the attribute, i.e. Maximum Amplitude Weighed Integrated Energy Spectra at dominant frequency, is capable of qualitatively image a complex thin channels structure better than seismic amplitude.

A new approach of stochastic seismic inversion has been developed in order to improve both accuracy and precision of thin-bed structure interpretation. The technique includes building a prior model utilizing apparent indicator from seismic data and constraining stochastic inversion with Maximum Amplitude Weighed Integrated Energy Spectra. Seismic amplitude and extracted wavelet is two main indicators which used to constrain the development of prior model. Constraining inversion with Maximum Amplitude Weighed Integrated Energy Spectra means utilizing the Maximum Amplitude Weighed Integrated Energy Spectra to control the 'degree of match' of model with the data. Bayesian theorem and Markov Chain Monte Carlo is side by side used to generate realization and sampling the generated realization, respectively. Twenty realizations have been generated from twenty simulations of twenty samples both for synthetic model and real seismic data. The result shows that the technique improves both accuracy and precision of thin bed interpretation. Improving accuracy affects to increasing resolution, while precision affects the reducing uncertainty of interpretation. Many reservoir characterization scenarios then can be built from the result of inversion.

Acknowledgment

Authors thank to UniversitiTeknologi PETRONAS for supporting the research.

References

[1] Cooke D. and Cant J., 2010, Model-based Seismic Inversion: Comparingdeterministic and probabilistic approaches, CSEG Recorder p 28-39

[2] Gunning, J. and Glinsky M. E., 2004, Delivery: an open-source model-based Bayesian seismic inversion program, Computers & Geosciences 30.

[3] Marangakis, A., J. K., Costain, and C. Coruh, Use of integrated energy spectra for thin-layer recognition: Geophysics, Vol. 50, No. 3, p. 495-500, 1985.

[4] Kallweith, R. S., and L. C. Wood, 1982, The limits of resolution of zero-phase wavelets: Geophysics, Vol. 47, No. 7, p. 1035-1046, 1982.

[5] Partyka, G.A., 2001, Seismic thickness estimation: three approaches, pros and cons: 70th Annual International Meeting Society of Exploration Geophysicists, Expanded Abstracts, 503-506.

[6] Taner, M.T., F. Koehler, and R. E. Sheriff, 1979, Complex seismic trace analysis: Geophysics, 44, 1041-1063.

[7] Ghosh D. P. et. al., 2007 Seismic Attributes add new dimension to prospect evaluation, SEG 2007 Denver

[8] Hardage, B. A., Levey R. A., Pendleton V., Simmons, J., and Edson R., 1994, A 3-D Seismic Case History Evaluating Fluvially Deposited Thin-Bed Reservoir in Gas-Producing Property: Geophysics, Vol. 59, No. 11, p. 1650-1665.

[9] Kerr, Dennis R., and Jirik, Lee A., 1990, Fluvial architecture and reservoir compartmentalization inthe Oligocene middle Frio formation, south Texas: Transactions of the Gulf Coast Association ofGeological Societies, Volume XL, p. 373–380.

Pore Type Classification on Carbonate Reservoir in Offshore Sarawak using Rock Physics Model and Rock Digital Images

L A Lubis[1] and Z Z T Harith[1]

[1] Geosciences Department, Universiti Teknologi PETRONAS, Perak, Malaysia

E-mail: luluan.lubis@gmail.com

Abstract. It has been recognized that carbonate reservoirs are one of the biggest sources of hydrocarbon. Clearly, the evaluation of these reservoirs is important and critical. For rigorous reservoir characterization and performance prediction from geophysical measurements, the exact interpretation of geophysical response of different carbonate pore types is crucial. Yet, the characterization of carbonate reservoir rocks is difficult due to their complex pore systems. The significant diagenesis process and complex depositional environment makes pore systems in carbonates far more complicated than in clastics. Therefore, it is difficult to establish rock physics model for carbonate rock type. In this paper, we evaluate the possible rock physics model of 20 core plugs of a Miocene carbonate platform in Central Luconia, Sarawak. The published laboratory data of this area were used as an input to create the carbonate rock physics models. The elastic properties were analyzed to examine the validity of an existing analytical carbonate rock physics model. We integrate the Xu-Payne Differential Effective Medium (DEM) Model and the elastic modulus which was simulated from a digital carbonate rock image using Finite Element Modeling. The results of this integration matched well for the separation of carbonate pore types and sonic P-wave velocity obtained from laboratory measurement. Thus, the results of this study show that the integration of rock digital image and theoretical rock physics might improve the elastic properties prediction and useful for more advance geophysical techniques (e.g. Seismic Inversion) of carbonate reservoir in Sarawak.

1. Introduction

Despite all the efforts and resources of new geological models, high resolution data and development of new technologies, the predictions of carbonate properties and fluid contents from seismic are still a difficult challenge. A great spatial variability of parameters e.g. porosity and permeability are the main problems that cause the prediction more complicated than in clastics. The Rock physical models that take into consideration of these specific carbonate rocks features can constrain quantitative seismic interpretation e.g. AVO and time lapse analysis approaches.

As distinct from sandstones, in carbonate systems, the combined effect of variations in depositional facies and diagenetic alteration plays a key role in controlling the variations in sonic velocities and acoustic impedance, which makes the seismic response of these rocks hard to interpret and understand [1]. Carbonate pore types are considered the key factor causing significant variations in permeability and elastic properties [5] [1]. Dolomitization, the process of calcium to magnesium substitution, is one of the most important processes of the reservoir properties degradation. Dolomitization usually

[1] L A Lubis.

decreases porosity [7] and increases bulk and shear moduli. Because of this complex behavior, in terms of geological models, it is important to establish the main trends.

The measurements from laboratory are crucial and importance for hydrocarbon exploration, especially in carbonate reservoir. Even the robust and sophisticated geological models which are formed from well logs and seismic datasets that improve the information for every year need to be confirmed and validated experimentally in the laboratory. The best way to achieve this goal is through laboratory experiments in controlled conditions. Advances in technology also make the laboratories better equipped and new valuable information can be obtained to help to understand the ultimate datasets. The number of laboratory ultrasonic measurements in carbonates will increase, and to address this demand, new and more efficient theoretical models need to be employed to explain these data. Finally, the understanding of the physical properties of carbonates, measured or estimated (but well calibrated), are fundamental for the quantitative seismic interpretation of the complex carbonate reservoirs.

The proposed research seeks to evaluate and analyze the published laboratory measurements data of elastic properties on reservoir carbonate samples in offshore Sarawak [1] and develop the appropriate workflows that can be possibly used in log analyses and quantitative seismic interpretation. Since the elastic properties of carbonates highly depend on the pore type and shape of the carbonate rocks [1] [5] [8] [9], we did a different approach to improve the rock physics model. We integrated the information of the microstructure of carbonate rock to the rock physics model. The microstructure of carbonate rock was downloaded from Andrea, 2013 [3] [4]. The microstructure is analyzed using high quality X-Ray micro-CT to construct realistic dual-nature porosity models, quantify stiffness and compliance porosities and extract elastic parameters for seismic waves. With this approach we maximize the use of laboratory analysis information, constrained by realistic geological information.

2. Methodology

2.1. Differential Effective Medium Model

The combined effect of variations in depositional facies and diagenetic alteration occurred during intense geological processes cause carbonate rocks to develop wide variations of pore types, such as interparticle, intercrystal, moldic, vuggy, intraframe and microcracks [7]. Although this classification is useful for characterizing petrophysical properties, relating these microstructures and trends to geophysical responses and further geophysical method application such as seismic inversion are very challenging. From previous geoscientist observation, normally, vuggy pores are very rigid and less affected by seismic wave propagation, while cracks are effective to reduce velocities in carbonates [5]. Xu and Payne, 2009 [2] demonstrated that Xu and White (1995) model for shaley sandstones can be extended to predict velocities in carbonate rocks. Following Xu and Payne's rock physics modeling approach for carbonates, we mixed minerals present in the rock using Voigt-Reuss-Hill averages. Differential Effective Medium (DEM) scheme is used to calculate the dry effective bulk and shear moduli for different geophysical pore types represented by the pore aspect ratio.

We abstractly represent seismic velocities in carbonate reservoir rocks by using three geophysical pore types as in Figure 1: (1) Reference pores, which mainly consist of inter-particle pores and is considered the dominant pore type in carbonate rocks, (2) Stiff pores, which represent moldic and vuggy pores, which are usually formed as a result of dissolved grains and fossil chambers. (3) Cracks, which can occur due to differential compaction, faulting and solution collapse in carbonates [6] [2] [9].

Each pore component is incrementally added to the rock matrix to calculate the elastic properties of the resultant effective medium. One versatile feature of the Xu-Payne model is that individual pore components can be included into the model such that they are either isolated, or, in perfect fluid connectivity with the remaining pore space. The pore space model is calibrated using a cross-plot of P-wave velocity against porosity [9].

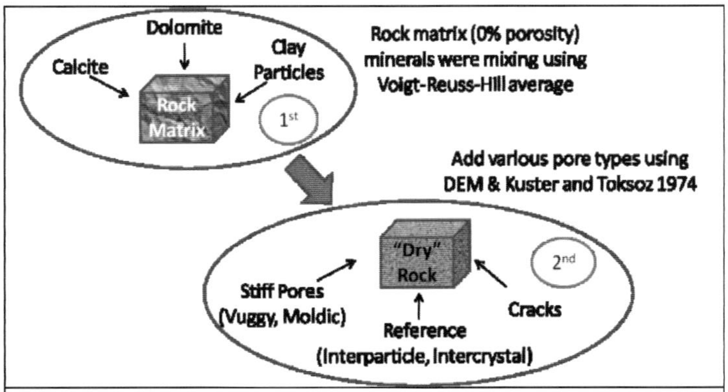

Figure 1. Diagram of DEM (modified from Xu et.al.[2]) to build carbonate rock physics model in our study.

Figure 1 shows how we create our carbonate rock physics model on dry rock (rock which not contained fluids). The first step is using a mixing law (the Voigt-Reuss-Hill average) to mixed the minerals which are present in the rock. The dominant minerals for carbonate rocks are usually calcite and dolomite. Last step is to incorporate differential effective media theory and Kuster-Toksoz (1974) theory to include type of pores (Moldic, Interparticle and Microcracks) and take into account the mechanical interaction between pores. The result of the calculation is the effective elastic properties of Dry Rock. Figure 2 below shows the predicted effect of pore type on P-wave velocity, the reference line for mixed minerals which contains interparticle and intercrystal pore types are also incorporated on the plot.

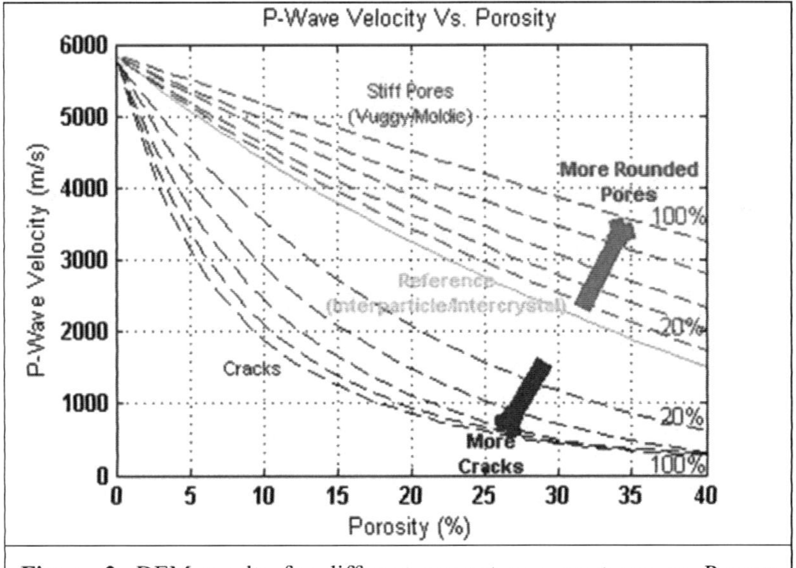

Figure 2. DEM results for different percentage pore types on P-wave velocity.

On the figure above the solid matrix was assumed with dominant mineral in calcite plus clay particles. The average Voigt-Reuss-Hill was used to calculate the minerals average of carbonate rocks for P-wave velocity. The reference represents a system with interparticle or intercrystal pore types. The curves below the reference curve represent systems with increasing percentage of cracks pore type. The curves above the reference curve represent systems with increasing percentage of stiff pore (vuggy / moldic).

2.2. Numerical Methods to Compute the Physical Properties
The numerical computations of the physical properties of rock were computed on a 3D CT-scan image of intercrystal carbonate rock. The data used in this proposed work was taken from a benchmark shared dataset from Andrea et.al, 2013 [3] [4]. The physical properties, e.g. porosity (total, connected and unconnected), absolute permeability and formation factor, were re-simulated and calculated using AVIZO commercial software. As for the calculation of the elastic properties were taken from Andrea et.al, 2013 [4] research work. The figure below shows the image for numerical computation of the physical properties.

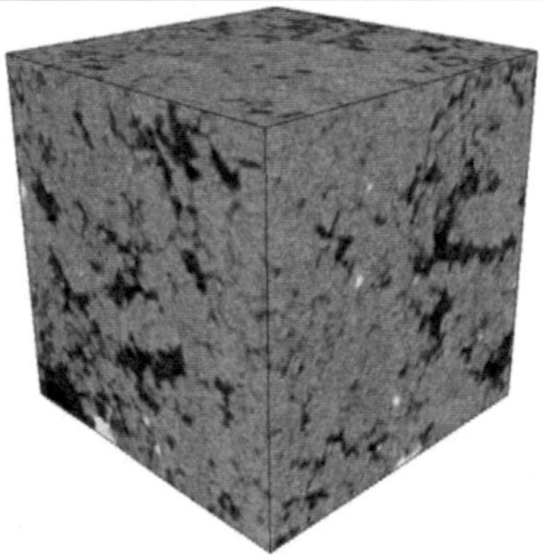

Figure 3. Intercrystalline Carbonate 400^3 pixels cube from micro-CT image [3].

2.2.1. Porosity
Three different porosities are reported: total porosity, percolating (connected) porosity and isolated (unconnected) porosity. Connected or percolating porosity is the fraction of total porosity that available to flow. The fraction of total porosity that unable for flow is categorized as unconnected or isolated porosity. The following relations hold for porosities of CT-Scan image:

$$\phi_{tot} = \phi_{perc} + \phi_{iso} \qquad (1)$$

2.2.2. Absolute Permeability
The Lattice Boltzmann Method (LBM) for fluid simulation is an appropriate choice for these cases. This method is a robust technique that simulates flow according to simple rules governing local interactions between individual particles and recovers the Navier-Stokes equations at the macroscopic scale.
Boltzmann equation solved by counting particle density distribution at time *t* and location of *r*. From the local flux, a volume averaged flux can be calculated. Then, the absolute permeability is computed in a manner analogous to a laboratory measurement: a pressure head or body force is directly applied to a digital sample. The resulting fluid flux is computed and permeability is calculated according to the Darcy's law.

$$k = \frac{\langle Q \rangle}{\nabla P}\mu \qquad (2)$$

The flow simulation is performed with pressure gradient (∇P) assigned across opposite faces of the 3D cube. Next, a volume averaged flux $\langle Q \rangle$ is computed from local flux. μ is the dynamic viscosity of the fluid.

2.2.3. Formation Factor
The steady state electrical conductivity, or formation resistivity factor (F), of a brine saturated rock is governed by the Laplace equation

$$\nabla \cdot \mathrm{J} = 0 \quad (3)$$

$$\mathrm{J} = \sigma_\omega \nabla \phi \quad (4)$$

subject to the boundary condition $\nabla\phi \cdot \mathrm{n} = 0$ on the solid walls (i.e. insulating walls). Here J is the electrical current, σ_ω is the electrical conductivity of the fluid that fills the pore space, Φ is the potential or voltage and **n** is the unit vector normal to the solid wall. Numerical solutions of the Laplace equation are obtained by a random walk algorithm or by a finite difference method.
The effective directional conductivities σ_i, $i = x, y, z$ are computed by applying a potential gradient across the sample in i-direction. The directional formation resistivity factor Fi is the inverse of the effective electrical conductivity $F_i = \sigma_i / \sigma_\omega$. We define the average formation resistivity factor F as the harmonic mean of direction dependent formation factors.

2.2.4. Elastic Properties
The finite element method described by Garboczi and Day (1995) has been implemented for calculation of elastic properties. The method uses a variational formulation of the linear elastic equations and finds the solution by minimizing the elastic energy using a fast conjugate-gradient method. The results are valid for quasi-static conditions or at frequencies which are sufficiently low such that the included pore pressures are in equilibrium throughout the pore space [10]. The effective bulk and shear moduli are computed assuming isotropic linear elastic behavior. The Vp and Vs are subsequently calculated using the simulated effective elastic moduli and the effective density according to:

$$V_p = \sqrt{\frac{K + \frac{4}{3}\mu}{\rho}} \quad (5)$$

$$V_s = \sqrt{\frac{\mu}{\rho}} \quad (6)$$

Inputs for the calculations are:
- A three-dimensional representation of the rock microstructure (a digital rock sample)
- Density (ρ) and elastic properties (K and μ) of each mineral composing the rock matrix
- Density (ρ) and elastic properties (K and μ) of the fluid present in the pore space.

Table 1. Density and elastic properties of dominant minerals in carbonate rocks [3] [4].

Mineral	Density (g/cc)	Bulk Modulus (GPA)	Bulk Modulus (GPA)
Calcite	2.71	68.3	28.4
Dolomite	2.87	94.9	45.7

3. Results and Discussion

The main objective of this study is to improve the elastic properties prediction on carbonate rock type by integrating pore type information, through rock physics (DEM) and rock digital image. If we successfully integrate pore types information with laboratory measurement and rock digital image, there is possibility to use seismic inversion on the Central Luconia carbonate area.

We will exam pore type effect on elastic properties and calibrate the pore geometry related parameters in effective medium based model then integrate with the effective elastic properties simulated on carbonate rock digital image. Figure 7 below shows the calibration of pore geometry with P-wave velocity from laboratory measurement.

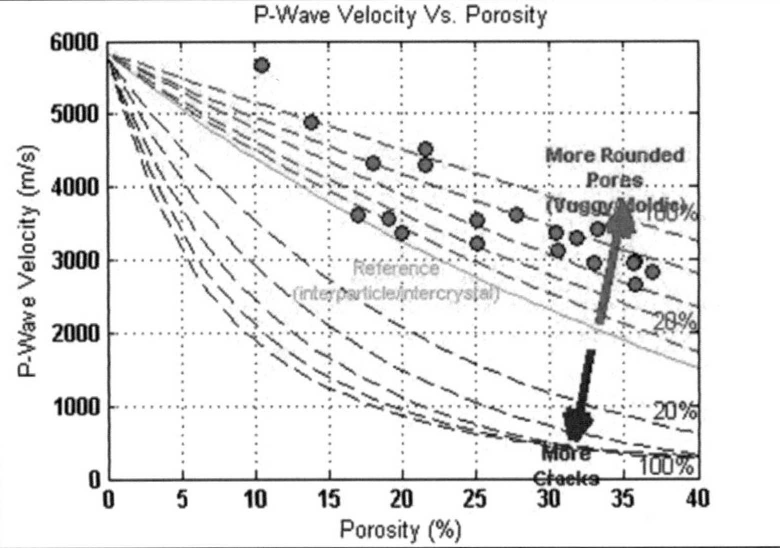

Figure 4. Illustration of possible pore type effects on the P-wave velocity and porosity relationship. All of the data points are sonic velocity data from [1].

Next step, we examined and integrate the results on the computation of physical properties on rock digital image. Table 2 below show the physical properties simulated on carbonate rock digital image.

Table 2. Simulation Results for Physical Properties on Carbonate Rock Digital Images.

Property		Image Volume	
Image Volume (voxel*)		400^3	1024^3
Porosity (%)	Con.	23.55	22.05
	Iso.	0.9	2.42
Bulk Modulus (GPa)		24.8**	25**
Shear Modulus (GPa)		14**	13.6**
Permeability (mD)		630.45	723
Formation Factor		26	30

*voxel = volume pixels, **Data from Andrea, 2013 [3] [4].

Calculated Vp are integrated on the previous plot and shown in Figure 5. Green line gives the best fit using Xu-Payne model with pore aspect ratio of 0.15. Red and blue dashed lines give the best fit with aspect ratio of 0.8 and 0.02, respectively. From the plot we can observe that the dominant pore types from 20 core plugs (grey dots) are stiff pores (vuggy/moldic). There are two data which, refer to red circles on Figure 5, align with the reference plot (interparticle/intercrystal) pore types.

The green squares refer to the simulation of effective P-Wave velocities which were calculated using the information on Table 2. Clearly, the compliance difference for the total porosity of 24.7%, and intercrystal pore type, the calculation of P-wave velocity is far above the reference line.

The fundamental reason of the difference was the input of the simulation which only uses pure calcite value for the simulation. As expected the results are faster than the reference line. The other aspect is there might not all the microstructures are resolved during the acquisition of CT-Scan data, since in carbonate the micropore/cracks might have the size of below 1 micron.

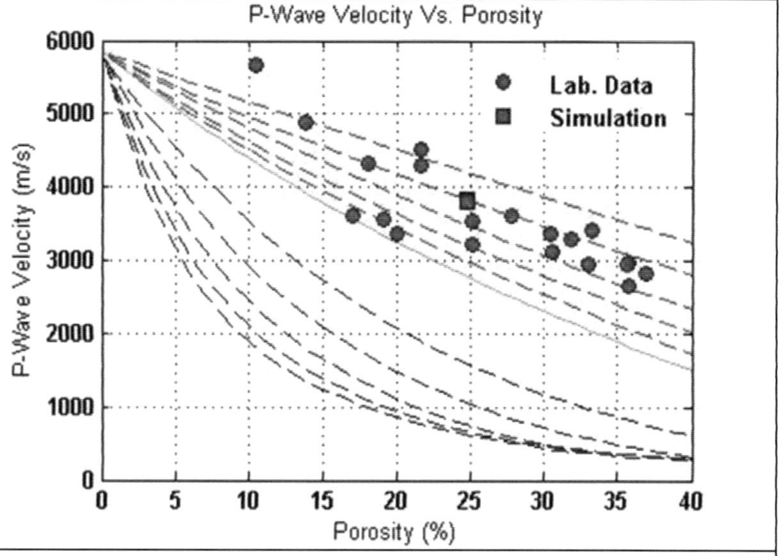

Figure 5. Green circles are P-Wave velocities computed on 3D CT scan image of carbonate digital image. Grey dots are P-Wave velocities from laboratory measurement [1]. Red and blue dashed line indicates vuggy give the best fit with aspect ratio of 0.8 and 0.02, respectively.

4. Conclusions

We have demonstrated that the integration of Rock digital image elastic properties simulation and Xu-Payne model is efficient for the use elastic properties and pore type classification prediction on carbonate rock. The possible of using Gassmann's theory for fluid substitution on carbonate might lead on the possible geophysical pore fluid inversion that can be calibrated on log and seismic data on carbonate reservoir, especially in Sarawak. We use the rock physics modelling approach that takes into account three defined geophysical pore types and shown the possibility to bridge them to the bigger scale such as log and seismic responses.

A more comprehensive study that includes more samples with various pore types and similar porosity values need to be carried out. This can be done (if the core samples are not available) on digital rock images of cuttings and sidewall core plugs which cannot be used to properly on the laboratory for measurement. It can be done fast and several times partition, as shown in this work, we show that the simulation on 1024 and 400 cubes on the same data.

This approach is practical, easily repeatable (in real time) and can be used as an alternative method when core plug is not available. The possibility to use this method on chips, cuttings and rotary sidewall cores that routinely available from wells is also align to obtain elastic properties distribution information since these samples give good sampling intervals which almost cover whole depth of the wells. However, there is always uncertainty associated with geophysical pore type inversion results as it depends on many different factors: the quality of seismic data, number of wells, input AI and porosity volumes, shale content, fluid content, etc.

References

[1] Bashah N S I and Pierson B 2012 Quantification of pore structure in a Miocene Carbonate Build-up Central Luconia, Sarawak and its relationship to sonic velocity *IPTC 2012*, Thailand

[2] Xu S and Payne M A 2009 Modeling elastic properties in carbonate rock *The Leading Edge* pp 66-74

[3] Andrea H et al 2013 Digital rock physics benchmarks - Part I: Imaging and segmentation *Computers & Geosciences* pp 25–32

[4] Andrea H et al 2013 Digital rock physics benchmarks - Part II: Computing effective properties *Computers & Geosciences* pp 33–43

[5] Eberli G P, Baechle G T, Anselmetti F S and Incze M L 2003 Factors Controlling Elastic Properties in Carbonate Sediments and Rocks *The Leading Edge* pp 665-660

[6] Lucia F J 1995 Rock-fabric/petrophysical classification of carbonate pore space for reservoir characterization *AAPG Bulletin* pp 1275–1300

[7] Lucia F J 1999 *Carbonate reservoir characterization* (Berlin, Springer-Verlag) p 226

[8] Weger R J, Baechle G T, Masaferro J L, and Eberli G P 2004 Effects of Porestructure on Sonic Velocity in Carbonates *Society of Exploration Geophysicists, Expanded Abstracts* pp 1774

[9] Kumar M, and Han D H 2005 Pore shape effect on elastic properties of carbonate rocks *Society of Exploration Geophysicists, Expanded Abstracts* pp 1477–1480

[10] Dvorkin J, Derzhi N, Diaz E, and Fang Q 2011 Relevance of computational rock physics *Geophysics*, pp 141-153

Assessment of hexavalent chromium release in Malaysian electric arc furnace steel slag for fertilizer usage

L K Bankole[1], S A Rezan[1] and N M Sharif[1]

[1]School of Materials and Mineral Resources Engineering, Universiti Sains Malaysia, 14300 Nibong Tebal, Pulau Pinang, Malaysia

E-mail : srsheikh@eng.usm.my

Abstract. This study investigates the leaching of hexavalent chromium (Cr (VI)) from electric arc furnace steel slag as Cr (VI) is classified as human carcinogen. Batch leaching tests were performed for 16 days. The lixiviants used were alkaline, de-ionized and rain water. After 16 days, Cr (VI) was found to be highest in alkaline water (0.03 mg/L) and lowest in de-ionized water (0.01 mg/L). Besides the lixiviants used, slag stirring speed and liquid to solid ratio also affect Cr (VI) released. The experimental work was complimented with slag characterization using XRF, XRD and SEM/EDX analysis. The leaching process was also simulated via Factsage software to calculate isothermal pourbaix diagrams. The Cr (VI) released was low and below the threshold of 0.1 mg/L set for public water systems. Recycle the slag as fertilizer should be considered safe as it does not exceed the safety limit set for Cr (VI) dissolution.

1. Introduction

Steelmaking in Malaysia is mostly produce using electric arc furnace (EAF) and about 8 million tons of steel is being consumed per year [1]. As a result of steel production, 10-15% of EAF steel slag is being generated as by-product [2]. This by-product contains valuable metals which have facilitated its utilization as aggregate in road and water-way construction, cement production, metallurgical fluxes and as soil amendment [3]. Currently in Malaysia, there is no utilization of EAF steel slag except being disposed in landfill [4]. The main concern is that slag recycling as soil amendment and fertilizer will pose a serious environmental problem due to the leachable heavy metals and contamination of the soil environs, surface and ground water.

Cr is one of the heavy metal contained in EAF steel slag and its presence may be due to the use of scrap metal with high Cr content in the EAF during steelmaking. Cr exists in the most stable form as Cr (III) and (VI) in the environment. Cr (III) is an essential element in human metabolism at low dose, relatively immobile and low in reactivity [5]. However, Cr (VI) is more toxic at moderate doses, highly mobile, soluble in water and classified as human carcinogen with negative health effects ranging from skin irritation to lung cancer [6-8]. The toxic effects of Cr (VI) are consequence of its readily penetrating into cell membrane and its strong oxidizing power. It forms a number of anions which are soluble and mobile in the environment [9]. They are weakly sorbed to soil under alkaline to slightly acidic condition leading to high mobility in the subsurface [10]. Thus, the maximum concentration of Cr (VI) allowed in drinking water is 0.1 mg/L due to Cr (VI) toxic nature and the potential oxidation of Cr (III) to (VI) [11].

Cr leaching have been studied extensively [3, 12-15], however, steel slag originating from various localities may differ in terms of leaching behaviour, mineralogy, extraction process and scraps charging process. In this study, modified European batch leaching test was employed to determine the release of Cr (VI) in Malaysian EAF steel slag. The leaching experiment was complemented with slag characterization and the leaching process also simulated via Factsage software using the EpH module. The results of this study will provide further understanding of Cr (VI) leaching from EAF steel slag and assist in further predicting the risk toxicity effect to the soil, surface and ground water.

2. Experimental

2.1. Slag sample characterization

EAF steel slag samples were ground in an agate mortar to avoid contamination and sieved to gain powder of less than 75μm. The -75μm powder were then sent for characterization. The chemical composition was determined by Rigaku XRF spectrometer, model RIX 3000. The mineralogical phases were identified using Bruker AXS diffractometer D8 Advance machine. The diffractometer were conducted at 40 mA and 40 kV using diffracted beam graphite-monochromatic with Cu radiation at a step size of 0.0340. The XRD patterns were measured in a 2θ range of 0–80° intervals and the analysis performed at room temperature. Qualitative analysis of XRD patterns were conducted using the Eva software and its database reference patterns. The morphology of the EAF slag was studied on polished sample using SEM coupled with EDX microprobe for in-situ chemical analysis. The SEM was a Zeiss Gemini Supra 35VP coupled with EDAX equipment.

2.2. Leaching test

Modified European batch leaching test EN 12457-2 [16] were performed for a period of 16 days. The lixiviant used were de-ionized (DI), alkaline and rain water. Particle sizes of the EAF slag was set at 1-4 mm. Each batch leaching experiment contained 50 g of EAF steel slag in bottles filled with the lixiviant solution corresponding to liquid solid (L/S) ratio of 20 at varying speed. Another set of leaching experiment were conducted at zero (0) rpm for 8 days at varying L/S ratio. Small samples of the leachate were taken every 24 hours and filtered through Whatman 42 filter paper. The filtered leachates were analyzed for Cr (VI) using UV-Vis spectrophotometer and the pH values measured using digital pH meter. XRD analysis were also performed on the residue slag samples after leaching for 30 days at 0 rpm with L/S ratio of 10 in DI water, to identify any new phases formed.

2.3. Simulation of leaching process

The EpH module was used to simulating the leaching process and plots isothermal pourbaix diagrams [17]. The leaching simulation parameters depend on the Cr (VI) ions concentration from the leaching experiment and slag chemical composition from the XRF analysis. The EpH module carries out calculations using pure stoichiometric substances and the simulation was performed at room temperature based on the FactSage database.

3. Results and Discussion

3.1. Slag sample characterization

The chemical composition of the EAF steel slag sample was shown in table 1. The slag mainly composed of Fe_2O_3, CaO, SiO_2, Al_2O_3, MgO, MnO, Cr_2O_3 and some minor metal oxide constituents. The basicity index of the slag calculated based on (CaO+MgO) / (SiO_2+Al_2O_3) works out to be 1.065. This basicity index has effect on the leaching of EAF steel slag due to dissolution and hydrolysis of basic oxides compare to the soluble acidic oxide. Besides, the chemical composition plays an important role in the leaching mechanism of Cr and the pH of the leachate [18]. The higher amounts of basic oxide in the slag samples, contributed to the slight increase in the pH values observed in solution. This gives high alkalinity and favourable to Cr (VI) leaching.

Table 1. Chemical composition of EAF steel slag by weight (%).

Slag	Fe_2O_3	CaO	SiO_2	Al_2O_3	MgO	MnO	Cr_2O_3	ZnO	TiO_2	P_2O_5	SO_3
EAF	33.3	27.5	19.3	9.40	3.07	3.55	1.54	0.80	0.57	0.49	0.42

The XRD analysis of the raw EAF steel slag and the leached samples were shown in Figure 1. The diffraction patterns were heterogeneous consisting of a mixture of crystalline phases. The analysis shows the presence of Calcium silicate oxide, Magnesium iron oxide, Wustite and Magnetite. Cr was not detected from the XRD as it was observed in the XRF and SEM/EDX analysis. This may be due to low intensity of their diffraction lines and the complexity of the diffractograms. Any other minor constituent phases were very difficult to assign as well. The XRD analyses after the leaching experiment indicate that the calcium silicate oxide phase was not dissolved in solution after the leaching for 30 days. This shows that this mineral phase was more tightly bonded together.

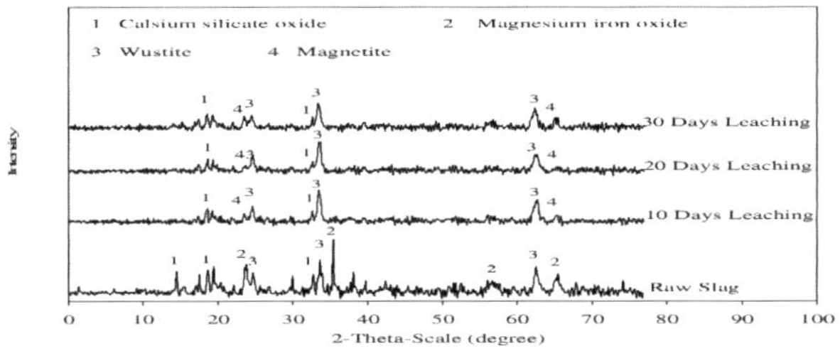

Figure 1. XRD pattern of EAF slag and leached samples for 30 days.

The SEM micrograph show distinct mineral phases as shown in figure 2. The presences of light and dark areas in this figure are explained by variable Fe, Ca, Si, Mg, Mn, Al and Cr contents as revealed by the EDX analysis. The EDX analyses revealed that light areas contained higher proportion of Fe, Mg, Mn, Cr and Al, while the dark area contained higher proportion of Ca and Si. Cr was associated within the light area as revealed by the EDX. The dissolution of this moderately-to-sparingly soluble phase (spinel phase) [19] will have effect on the leaching of the Cr from the EAF steel slag.

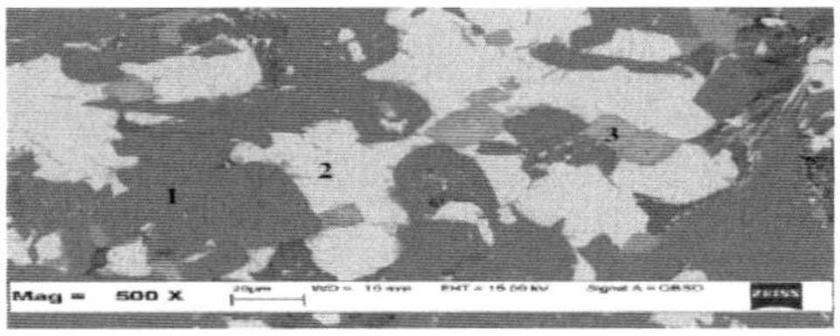

Figure 2. SEM micrograph of EAF slag sample.

The EDX results also confirm the presence of minor constituent phases such as titanium already observed by XRF spectrometer. With the coupled EDX, the EAF steel slag shows predominant presence of Fe, Ca, Si, Mg, Mn, Al and Cr. The EDX points 1, 2 and 3 are shown in table 2.

Table 2. EDX point 1, 2 and 3 of EAF slag sample by weight [%].

EDX point	O	Mg	Al	Si	Ca	Ti	Cr	Mn	Fe
Point 1	27.04	-	-	18.68	50.36	-	-	-	-
Point 2	16.64	4.13	0.51	0.22	2.33	0.50	1.64	9.17	61.93
Point 3	24.85	5.15	5.33	-	-	-	42.61	-	18.53

3.2. Leaching test

3.2.1. Effect of different lixiviant. Figure 3 shows the effect of different lixiviant on Cr (VI) leaching at 0 rpm. After 16 days, Cr (VI) was found to be highest in alkaline water (0.03 mg/L) and lowest in DI water (0.01 mg/L). In alkaline water, the pH was high, thus Cr (VI) was steadily leached. After 13 days, the leaching chemical reaction still continues with the alkaline water due to slow rate of initial reaction. Besides, Cr (VI) was favourable to form at high pH as shown in the pourbaix diagram (Figures 9 and 10), which also corroborate the leaching result. The pH reading in the different lixiviant shows similar trend but the DI water has the lowest pH value as shown in Figure 4. The leachates pH was alkaline and these were generated within the first 24 hours of contact with the lixiviant.

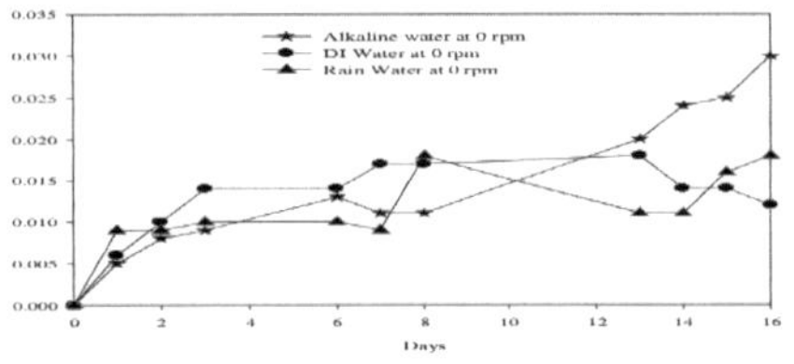

Figure 3. Effect of lixiviant on Cr (VI) leaching.

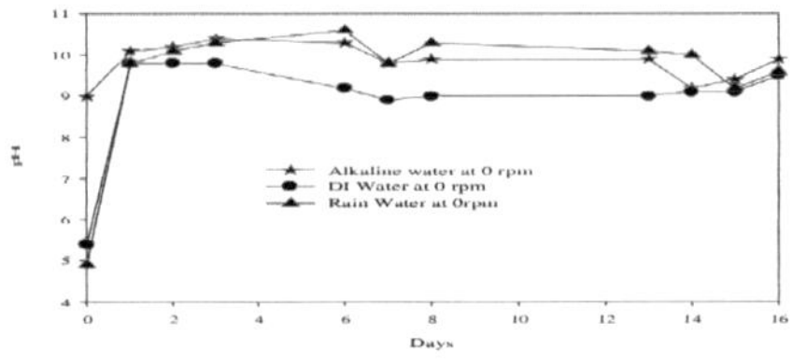

Figure 4. Effect of lixiviant on leachate pH.

The increases in the amount of Cr (VI) leached and later decreased, could be interpreted as the reduction of Cr (VI) by precipitation of reductant mineral. Fe (II) could act as a reductant to Cr (VI) and more soluble acidic oxide will also oxidize Fe (II) to Fe (III), leading to a lower potential for Cr (VI) reduction, thus more Cr (VI) will remains in solution [6, 20, 21]. Cr (VI) leaching is thereby related to redox reactions taken place in the leaching process.

3.2.2. Effect of stirring speed. Figure 5 shows the effect of stirring speed on Cr (VI) leaching using DI water. Cr (VI) was found to be highest at 50 rpm within the first seven days. Increase in stirring speed leads to an increase in the concentration of Cr (VI) leached from 0.006 mg/L at 0 rpm to 0.032 mg/L at 50 rpm within the first 24 hours. The pH at 30 and 50 rpm shows similar trend and both falls within the same values as shown in Figure 6. As the speed was increasing, there was more contact between the slag samples and the lixiviant. It was found that the highest concentration of Cr (VI) was leached within the first 24 hours at 30 and 50 rpm. Thus, increase in stirring speed indicates that the release of Cr (VI) may be controlled by diffusion.

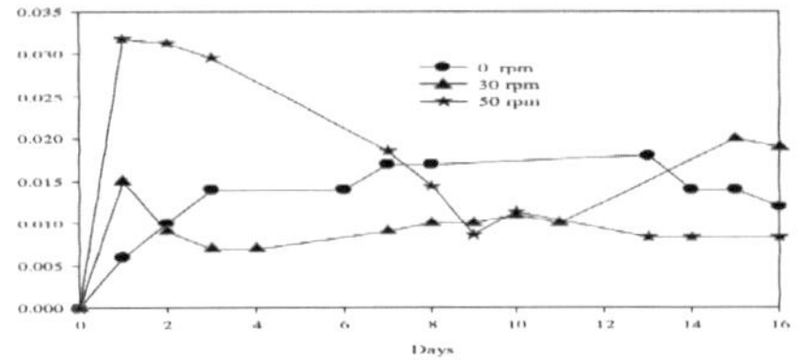

Figure 5. Effect of stirring speed on Cr (VI) leaching.

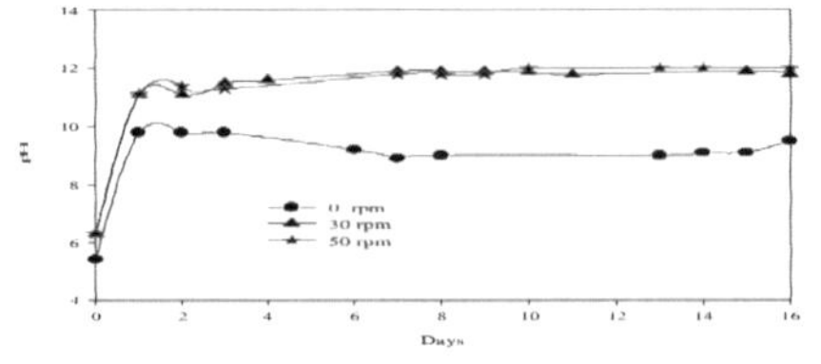

Figure 6. Effect of stirring speed on pH of the leachate.

3.2.3. Effect of liquid to solid ratio. Figure 7 shows the effect of L/S ratio on Cr (VI) leaching at 0 rpm using DI water. Cr (VI) was found to be highest at L/S ratio of 5. The concentration of Cr (VI) leached increases as the L/S ratio decrease. At L/S ratio 20, 10 and 5, the concentration of Cr (VI) leached increases from 0 to 0.017, 0.023 and 0.032 mg/L respectively at the end of the leaching experiments. The highest pH was also recorded when the L/S ratio was 5 within the first 48 hours of the leaching experiment as shown in Figure 8. The volume of the lixiviant could affect the concentration of leached slag into the solution due to the dilution factor.

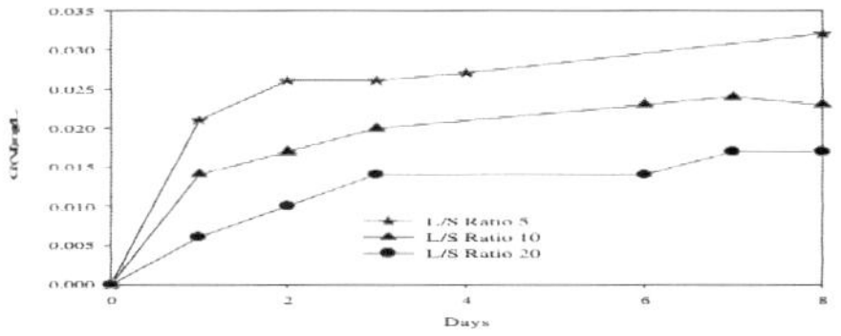

Figure 7. Effect of liquid to solid ratio on Cr (VI) leaching.

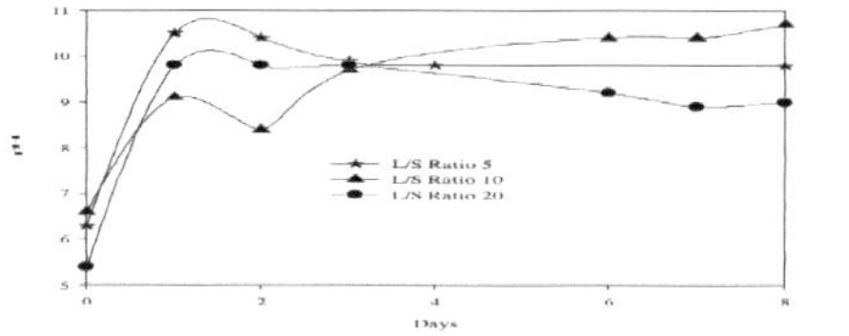

Figure 8. Effect of liquid to solid ratio on pH of the leachate.

3.3. Simulation of leaching process

The simulated leaching process for slag system predicted was shown in Figures 9 and 10. From the SEM/EDX analysis, substantial fraction of the Cr content was contained in the phase detected by XRD as magnesium iron oxide. This phase was regarded as a possible source of the Cr leaching. Pourbaix diagram for the Fe-Mg-Cr-H_2O and Ca-Si-Cr-H_2O system containing 2 wt % Cr at 25°C was analyzed to verify the most suitable leaching condition for the Cr to dissolves into solution. The concentration of Cr (VI) ions in the leachate reaches 0.03 mg/L ($0.6e^{-6}$ mol/Kg) from the leaching experiment. This value was used in the simulation and weight percent ratio of the elements contained in the EAF slag.

The simulation shows that Cr (VI) species ($HCrO_4^-$) is thermodynamically favourable when the redox potential (Eh) is within 0.7 to 1.4 at acidic pH range. In the neutral and caustic solutions, Cr_2O_3 is stable near zero potential. It dissolves as chromate (CrO_4^{2-}) in anode conditions and as the solution becomes more caustic in the cathode conditions. Under oxidizing conditions, Cr (III) will be converted into Cr (VI), resulting in the toxic and more mobile form of Cr. Thus stabilization of aqueous Cr (III) requires strong reductant concentration. From the pourbaix diagram, Cr (VI) is favourable to form at any environmental condition at higher redox potential and when the solution is more caustic. At high potential (Eh > 0), the system tends to remove electrons from the species (oxidizing condition), and at low potential (Eh < 0), the system is able to supply electrons to the species (reducing conditions). Thus, redox potential is a measure of the tendency of a chemical species to gain or loss electrons.

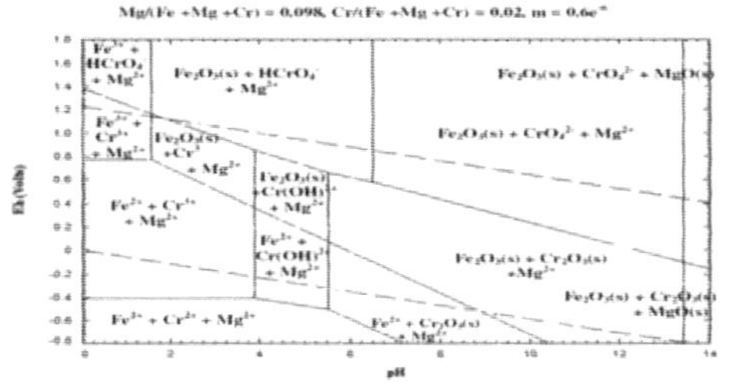

Figure 9. Pourbaix diagram of Fe-Mg-Cr-H_2O system containing 2 wt % Cr.

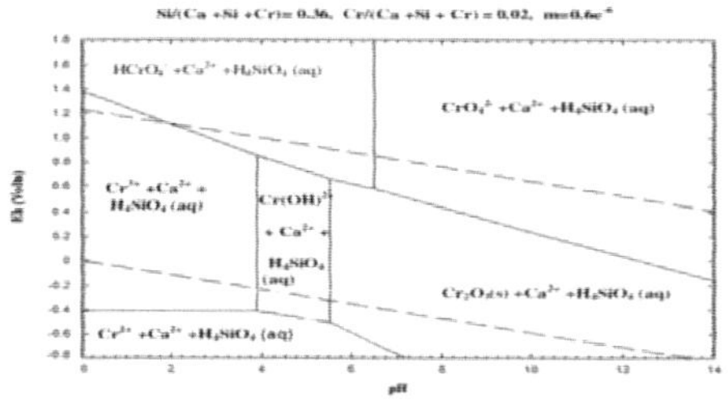

Figure 10. Pourbaix diagram of Ca-Si-Cr-H_2O system containing 2 wt % Cr.

4. Conclusion

Cr (VI) leaching was found to be highest in alkaline water and lowest in de-ionized water. Besides the different lixiviant used, increase in stirring speed leads to an increase in the concentration of Cr (VI) leached from 0.006 mg/L at 0 rpm to 0.032 mg/L at 50 rpm within the first 24 hours. Also, the concentration of Cr (VI) leached increases as the L/S ratio decrease. At L/S ratio 20, 10 and 5, the concentration of Cr (VI) leached increases from 0 to 0.017, 0.023 and 0.032 mg/L respectively at the end of the leaching experiments. The slag characterization shows that Cr (VI) leaching was related to the mineral components bounded within stable mineral phases detected by XRD as magnesium iron oxide, where Cr was situated as observed in the EDX analysis, and also the relative amounts of the soluble basic and acidic oxides in the EAF slag. From the pourbaix diagrams, Cr (VI) is favourable to formed when the solution is more caustic and at higher redox potential. Slag recycling as fertilizer usage in the environment does not have any negative effect and is safe, as the maximum Cr (VI) concentration released from the leaching experiment was low and does not exceed the safety limit set for Cr (VI) dissolution in the soil environs, surface and ground water.

Acknowledgement

The authors' wishes to thank Institute of Postgraduate Studies (IPS), Universiti Sains Malaysia for the USM fellowship, RU-PRGS 1001/PBAHAN/8045015 grant and also, support from USM short term grant 60312002 provided in the course of this research work.

References

[1] SEAISI 2012 SEAISI Statistical Year Book *Overview of Steel Consumption Pattern (SEAISI Newslatter July Edition)* ed P. Tepa-Apirak (Malaysia: SEAISI) p. 14

[2] Reddy AS, Pradhan RK and Chandra S 2006 Utilization of basic oxygen (BOF) slag in the production of a hydraulic cement binder *Int. J. Miner Process* **79** 98-105

[3] Proctor DM, et al 2000 Physical and chemical characteristics of blast furnace, basic oxygen furnace, and electric arc furnace steel industry slags *Environ Sci. Technol* **34**(8) 1576-82

[4] Lion Group 2010 Lion industries in tripartite venture for RM30mil slag processing plant *The Star Online Business News* Petalng Jaya, Malaysia

[5] Higgins TE, Halloran AR, Dobbins ME and Pittignano AJ 1998 In situ reduction of hexavalent chromium in alkaline soils enriched with chromite ore processing residue *J. Air & Waste Manage Assoc* **48**(11) 1100-6

[6] Cohen B, Lewis A, Petersen J, Von Blottnitz H, Drews S and Mahote S 1999 The TCLP and its applicability for the characterisation of worst case leaching of wastes from mining and metallurgical operations *Adv Environ Res* **3**(2) 152-165

[7] Sheps-Pelleg S and Cohen H 1999 Evaluation of the leaching potential of trace elements from coal ash to the (Ground water) aquifer *Int Ash utilization Symp* (University of Kentucky)

[8] Wazne M, Jagupilla SC, Moon DH, Christodoulatos C and Koutsospyros A 2008 Leaching mechanisms of Cr(VI) from chromite ore processing residue *J. Environ. Qual.* **37** 2125–34

[9] Saleh FY, Parkerton TF, Lewis RV, Huang JH, and Dickson KL 1989 Kinetics of chromium transformations in the environment *Sci. The Total Environ.* **86** 25-41

[10] Wittbrodt PR and Palmer CD 1995 Reduction of Cr (VI) in the presence of excess soil fulvic acid *Environ. Sci. & Technol.* **29**(1) 255-263

[11] USEPA 2009 National primary drinking water regulations *Available from: http://water.epa.gov/drink/contaminants/upload/mcl-2.pdf.*

[12] Facchinelli A, Sacchi E, Tararbra M and De Luca DA 2001 Assessment of chromium release from steel manufacturing waste *Environ. Geochem. Hlth* **23** 235-240

[13] Weng CG, Huang CP and Sanders PF 2001 Effect of pH on Cr(VI) leaching from soil enriched in chromate ore processing residue *Environ. Geochem. Hlth* **23** 207-211

[14] Gomes JF and Pinto CG 2006 Leaching of heavy metals from steelmaking slags *Revista Metalurgia Madrid* **42**(6) 409-416

[15] Fällman A 2000 Leaching of chromium and barium from steel slag in laboratory and field tests-a solubility controlled process? *Waste Manage* **20**(2-3) 149-154

[16] Vaajasaari K 2005 Leaching and ecotoxicity tests as methods for classification and assessment of environmental hazard of Solid Wastes *Tampere University of Technology* Publication **540**

[17] Bale CW, Chartrand P, Degterov SA, Eriksson G, Hack K, Mahfound RB, Melancon J, Pelton AD and Peterson S 2002 FactSage thermochemical software and database *Calphad* **26**(2) 189-228

[18] Gitari WM, Fatoba OO, Petrik LF and Vadapalli VRK 2009 Leaching characteristics of selected South African fly ashes: Effect of pH on the release of major and trace species *J. Environ. Sci. Hlth.* Part A **44**(2) 206-220

[19] Mostaghel S, Engstrom F, Samuelsson C and Bjorkman B 2010 Stability of spinel in a high basicity EAF slag *6th European Slag Conference* Madrid, Spain.

[20] Buerge IJ and Hug SJ 1999 Influence of mineral surfaces on chromium(VI) reduction by iron(II) *Environ. Sci. Technol.* **33**(23) 4285-91

[21] Weng CH, Huang CP, Allen HE, Cheng AHD and Sanders PF 1994 Chromium leaching behavior in soil derived from chromite ore processing waste *Sci. The Total Environ.* **154**(1) 71-86

The Attribute for Hydrocarbon Prediction Based on Attenuation

Maman Hermana[1], Z.Z.T.Harith, C.W.Sum, D.P.Ghosh

Geosciences Department, Universiti Teknologi Petronas
Seri Iskandar,Perak-Malaysia

mherjody@yahoo.com

Abstract. Hydrocarbon prediction is a crucial issue in the oil and gas industry. Currently, the prediction of pore fluid and lithology are based on amplitude interpretation which has the potential to produce pitfalls in certain conditions of reservoir. Motivated by this fact, this work is directed to find out other attributes that can be used to reduce the pitfalls in the amplitude interpretation. Some seismic attributes were examined and studies showed that the attenuation attribute is a better attribute for hydrocarbon prediction. Theoretically, the attenuation mechanism of wave propagation is associated with the movement of fluid in the pore; hence the existence of hydrocarbon in the pore will be represented by attenuation attribute directly. In this paper we evaluated the feasibility of the quality factor ratio of P-wave and S-wave (Qp/Qs) as hydrocarbon indicator using well data and also we developed a new attribute based on attenuation for hydrocarbon prediction -- Normalized Energy Reduction Stack (NERS). To achieve these goals, this work was divided into 3 main parts; estimating the Qp/Qs on well log data, testing the new attribute in the synthetic data and applying the new attribute on real data in Malay Basin data. The result show that the Qp/Qs is better than Poisson's ratio and Lamda over Mu as hydrocarbon indicator. The curve, trend analysis and contrast of Qp/Qs is more powerful at distinguishing pore fluid than Poisson ratio and Lamda over Mu. The NERS attribute was successful in distinguishing the hydrocarbon from brine on synthetic data. Applying this attribute on real data on Malay basin, the NERS attribute is qualitatively conformable with the structure and location where the gas is predicted. The quantitative interpretation of this attribute for hydrocarbon prediction needs to be investigated further.

1. Introduction

The main challenge in the seismic exploration is to predict hydrocarbon prospecting. Many seismic attribute have been proposed such as fluid factor (Smith, 1996), orthogonal Lame Constant (Goodway, Chen, & Downton, 1996), Poisson ratio (Verm & Hiltemmn, 1995) and AVO rotation (Connolly, 1999) and others. However, these attributes were based on the amplitude analysis. Interpretation based on amplitude still has an ambiguity because of the amplitude as seismic response is determined by many factors such as wave field divergence or spreading, angle dependent reflection and transmission losses, angle dependent ghosting in marine situation, inelastic absorption, internal peg-leg multiple energy loss, scattering, and mode of wave conversion. The interpretation based on amplitude is non-unique; it means that there are a lot of conditions that can produce the same seismic responses.

In case of Malay basin field, a poor reservoir containing gas may have the same amplitude response as relatively better quality sand with an oil pore fluid. Also the highly porous sand with brine could yield a bright spot as well as gas sand. This ambiguity remains an added complication in AVO and amplitude analysis and several false anomalies have been drilled. The lithology and fluid effect is hard to be distinguished in amplitude analysis. Hence there is a need to find out a new attribute that can representing only pore fluid rather than combination both of lithology and pore fluid effect.

Indication of attenuation mechanism as hydrocarbon indicator has been reported by some authors. Castagna et. al [1] reported that there is shadow zone beneath gas reservoir, they used the Instantaneous Spectral Analysis (ISA) method to detect the low frequency shadow. Korneev et al [2] explain the low frequency is associate with low quality factor of medium due to the fluid flow which is agree with diffusive-viscous theory. They used the frequency-dependent amplitude and phase-reflection properties to detect and monitor fluid saturated layers.

Effect of fluid flow in the pore during elastic waves propagation has been predicted analytically in the Biot's theory for low and high frequency [3]. The squirt flow effect of unbounded fluid during wave propagation will produce new wave field that can absorb a part of wave energy irreversibly. As consequence, the transmitted wave will be attenuated significantly. Not only P wave, but also S-wave undergoes the attenuation when it is propagated in the dispersive medium. The degree of attenuation in the medium is representing in the quality factor of P wave (Qp) and S-wave (Qs).

This paper discusses the possibility of Qp/Qs for hydrocarbon prediction especially in the Malay Basin field. For comparison, the Qp/Qs was compared with the Lamda over Mu and Poisson ratio and the performance of them in determining the hydrocarbon have been examined. In the other part of this work, a new attribute for hydrocarbon prediction which is based on attenuation has been developed and it was tested on synthetic and real data.

2. Background theory

2.1. Attenuation mechanism and definition

In the damped harmonic oscillation theory, the solution of wave propagation is represented by a complex number in wave number, where the imaginary part of that complex number is representing the attenuation. The solution of damped harmonic oscillation is:

$$A(x,t) = A_0 e^{i(k'x-\omega t)} \tag{1}$$

where the A is amplitude with complex wavenumber $k'=k+i\alpha$, hence the equation (1) can be formulated as:

$$A(x,t) = A_0 e^{-\alpha x} e^{i(kx-\omega t)} \tag{2}$$

In equation (2), we can see that amplitude decay by α factor. The α is defined as the absorption or attenuation coefficient. In term of energy, the attenuation coefficient is defined as loss energy per cycle of wavelength. By defining the quality factor (Q), attenuation can be formulated as: (Toksoz, Johnston, Seismic Wave Attenuation, SEG publication),1981.

$$\frac{1}{Q(\omega)} = \frac{1}{\pi}\frac{A_0(1-e^{-\alpha\lambda})}{A_0} \cong \frac{\alpha\lambda}{\pi} \quad or \quad Q(\omega) \cong \frac{\pi}{\alpha\lambda} \tag{3}$$

Equation (3) is showing that the attenuation of medium can be represented in term of quality factor (Q). The term of attenuation in this context is associated with intrinsic attenuation which is properties of medium.

2.2. Attenuation and rock physics relation

The attenuation and phase velocity of plane wave propagation in the viscoelastic medium is govern by Kramers-Kronig relation [4] and maximum of quality factor can be estimated using equation:

$$2Q_p^{-1} = \frac{M_\infty - M_0}{\sqrt{M_0 M_\infty}}$$
$$2Q_s^{-1} = \frac{G_\infty - G_0}{\sqrt{G_0 G_\infty}} \tag{4}$$

where the M and G are compressional and shear modulus, the indexing is representing the relaxed and unrelaxed condition. The equation (4) shows that inverse quality factor can be predicted from high and low frequency condition. By adopting the Hudson crack theory for isotropic model the Qp/Qs is determined as[4]:

$$\frac{Q_P^{-1}}{Q_S^{-1}} = \frac{5}{4}\frac{(M/G-2)^2}{(M/G-1)} / [\frac{2M/G}{(3M/G-2)} + \frac{M/G}{3(M/G-1)}] \tag{5}$$

3. Methodology

The main objective of this study is to find out a new attribute for hydrocarbon detection based on attenuation mechanism. In general, this work is divided into two part; feasibility study of Qp/Qs as hydrocarbon indicator in the well data and the development of new attribute based on attenuation for hydrocarbon prediction purposes.

3.1. Feasibility study on well data

In the first part, a Qp/Qs is estimated from well data using equation (5), where the M and G are derived from well log. To correlate this Qp/Qs parameter with hydrocarbon content, a well data from Malay Basin field has been examined. Crossplot, trend analysis and the contrast have been analysed to showing the performance of Qp/Qs in determining the hydrocarbon. Finally, the performances of this parameter were compared with the existing fluid indicator; Poisson's ratio and Lamda over Mu.

3.2. Normalized Energy Reduction Stack

In the second part, the new attribute Normalized Energy Reduction Stack (NERS) was defined. This attribute are operated on gather data to get the stack of energy reduction due to the attenuation. Assumption was made that in the gathers, the spectral energy of far offset will be reduced more than the near offset if the medium contain the hydrocarbon. Hence the stacking over gather of this energy reduction will represent the attention. If the existence of hydrocarbon in the pore is directly can be represented by attenuation, the NERS attribute hopefully can be used as direct hydrocarbon indicator. The gather with hydrocarbon will has high NERS and vice versa. Normalized are taken by divide the spectra with its maximum value, this technique are performed to minimize the effect of amplitude hence the different lithology will not affect too much on this attribute.

Mathematically the NERS attribute is formulated as:

$$NERS = \sum_j \frac{\left[(A_{ref}(f)^2 - A_{av,ref}) - (A_j^2(f)^2 - A_{av,j})\right] . f}{\sum f} \tag{6}$$

where the amplitude spectral are normalized using the maximum value in each windows target. Schematic of this definition is illustrated in the Figure 1 :

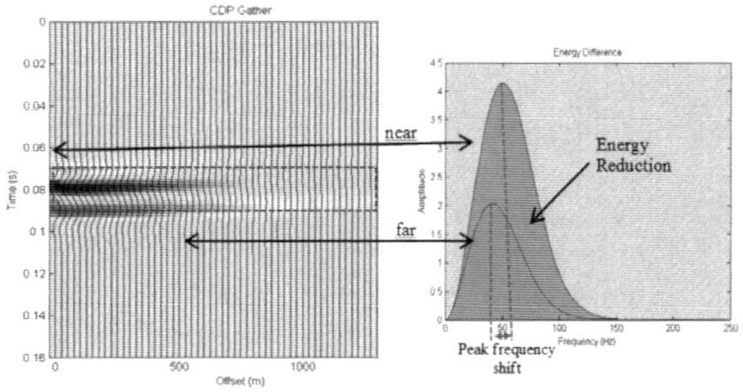

Figure 1. Illustration of Energy reduction.

The steps to calculate NERS from gather consist of; windowing of reflection target, performing the time frequency analysis, normalizing the spectral using its maximum value, calculating the integral of the spectra over all frequency, subtraction the value for every trace by zero offset trace and finally the summation all the residual for all offset are calculated. Before this technique is applied into real data, this method was tested using synthetic data using AVO class 2 model. In the final testing, a set of partial stack data from Malay Basin field was used to test the NERS attribute.

4. Result and discussion

4.1. Feasibility of Qp/Qs as hydrocarbon indicator in Malay Basin Field

In well domain, the values of Qp/Qs were calculated based on Vp, Vs and density logs. The curve is shown in **Figure 2**a. Compared with water saturation log, the Qp/Qs curve is tie similarly. High value of Qp/Qs associates with low water saturation log and vice versa. This result showing that the Qp/Qs is effectively can be associated with hydrocarbon saturation. Comparing with other pore fluid indicator (Lamda over Mu and Poisson ratio), the Qp/Qs has highest correlation value with water saturation curve. The coal bed formation is shown as low value in Qp/Qs log unlike in Lamda over mu and Poisson ratio curve. This fact indicated that coal will not be identified as hydrocarbon on Qp/Qs analysis. The ambiguity between coal and gas sand will be solved in the Qp/Qs analysis.

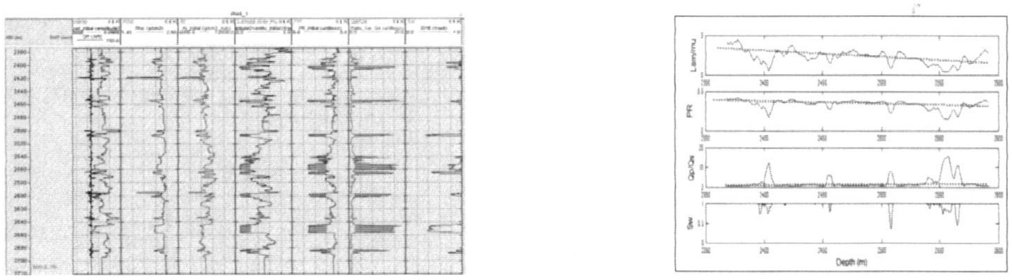

(a) (b)

Figure 2. (a) Qp/Qs curves compared with Lamda over Mu, Poisson ratio and water saturation, and its trends (b)

Trend analysis of Qp/Qs is performed by plotting the Qp/Qs value over depth and it's spectral. Using the water saturation trend as reference, the Qp/Qs trend was compared with Lamda over Mu and Poisson ratio trend, the result shown that Qp/Qs trend is more close to water saturation trend rather than Lamda over Mu and Poisson ratio trend as shown in the **Error! Reference source not found.**2b. In this figure, we can see that the Lamda over mu and Poisson ratio still has trend over depth not like Qp/Qs trend; it might be that both of them still effected by other factor such as compaction and

IOP Conf. Series: Earth and Environmental Science **19** (2014) 012005 doi:10.1088/1755-1315/19/1/012005

porosity that is embedded as lithology factor. In the spectral analysis also, the Qp/Qs spectral is more similar with water saturation spectral than Lamda over Mu and Poisson ratio.

Fluid content detectability using Qp/Qs was evaluated on Malay basin data. This parameter was compared with Lamda over mu and Poisson ratio. In the various scenarios of reservoir models, the Qp/Qs contrast in the gas sand is very high compared with Lamda over Mu and Poisson ratio. The effect of lithology on Poisson ratio and Lamda over Mu is still strong but not on Qp/Qs. In case of Gas-soft sand overlay by soft or hard shale, the Qp/Qs contrast is still highest compared with Lamda over mu and Poisson ratio. From this analysis, it can be concluded that the Qp/Qs is more powerful than Lamda over Mu and Poisson ratio for hydrocarbon detection.

4.2. Normalized Energy Reduction Stack (NERS) as Hydrocarbon Indicator

To investigate the ability of NERS method in discriminating the lithology and fluid, this method was applied into synthetic data. The synthetic model (velocity and density) were used here is the data from Malay basin field, meanwhile the Q factor is adopted from literature. The models consist of simple layering model (shale-sand-shale) with different fluid content is shown in the **Figure 3**. The thickness of sand bed is same for different fluid type. The wavelet source is used has dominant frequency 30 Hz.

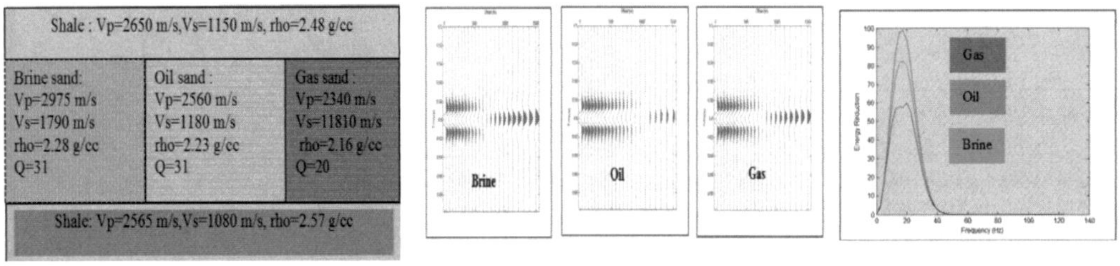

(a) (b) (c)

Figure 3. (a) Simple 3-layers model (brine, oil and gas saturated), **(b)** AVO response and **(c)** NERS before stacked response.

Seismic response was generated using the convolution model, the attenuation filter applied in the frequency domain. The NMO effect was neglected. The seismic response and the NERS calculated from synthetic are shown in the **Error! Reference source not found.**3a and 3b. This **Error! Reference source not found.**3 is showing that NERS attribute can be used to detect the fluid. Gas sand has highest NERS than Oil and brine sand. Oil still can be separated from brine (NERS of oil sand is larger than brine sand).

4.3 Implementation of NERS on Real Data

The seismic data from Malay Basin was used to implement The NERS analysis. Data consist of near, mid and far stack. Geologically, this area have anticline structure where is the gas potential is located in the top of these structure. The near, mid and far stack data are shown in the **Figure** 4.

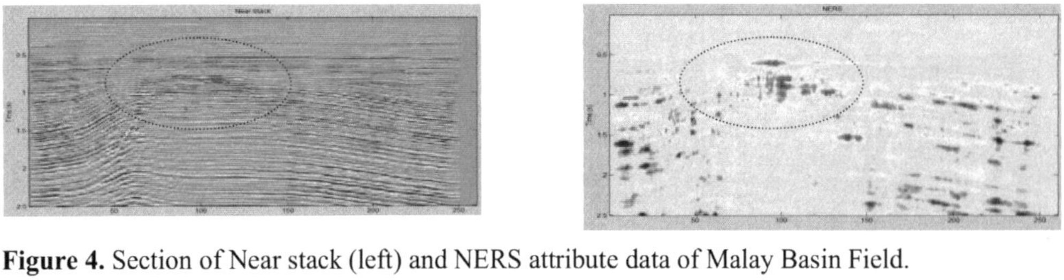

Figure 4. Section of Near stack (left) and NERS attribute data of Malay Basin Field.

The attribute NERS was calculated from these section using S-transform. The result is shown in the **Figure 4 (right)**. The Gas area is very clear shown in NERS attribute (see the circle). The Gas area is representing with high NERS value (red=high, blue =low). Others gas pocket is shown as spot which are located at A and B location. These spot is not shown in the amplitude on partial stack section. This attribute is conformable with the structure which has closure area. This area is probably the prospect of gas area.

5. Conclusion

Based on the analysis on the Qp/Qs behavior, the Qp/Qs curve that was derived from elastic properties is similar with water saturation or hydrocarbon saturation curve, hence the Qp/Qs is potential to be used as hydrocarbon indicator. Applied this attribute on Malay basin data, the result showing that the Qp/Qs is potential to be used for hydrocarbon detection. On the trend and contrast analysis, the Qp/Qs also had shown the trend close to water saturation trend. It can be concluded that the Qp/Qs able to be correlated with water saturation. In the contrast test, the contrast of Qp/Qs attribute is highest than other hydrocarbon attribute such as Lamda over Mu and Poisson ratio. The Qp/Qs contrast is highest in all soft/hard shale-soft/hard sand interfacing for hydrocarbon pore filled condition then Lamda over Mu and Poisson ratio contrast. This study shown that the Qp/Qs is more sensitive with pore fluid rather than the lithology changes.

The implementation of the NERS attribute on synthetic study and real data shown that this attribute which is derived based on attenuation is able to distinguish hydrocarbon from brine. The hydrocarbon has high NERS value mean while the brine has low NERS value. The difference is significant to distinguish both of them. Implementation on real data, the result shown that the NERS attribute is able to indicate the hydrocarbon prospect. The suspected hydrocarbon area was indicated by this attribute. Hence the NERS attribute is potential to be used for hydrocarbon prediction, however the quantitative interpretation of this attribute still need further investigation.

Acknowledgment

We thank UTP for funding this research under Graduate Assistant schema and Ikon Science for Donating Rockdoc Software.

References

[1] J. P. Castagna, S. Shengjie, and R. W. Siegfried, "Instantaneous spectral analysis : Detection of low-frequency shadows associated with hydrocarbons," *The Leading Edge*, pp. 120–127, 2003.

[2] V. A. Korneev, G. M. Goloshubin, T. M. Daley, and D. B. Silin, "Seismic low-frequency effects in monitoring fluid-saturated reservoirs," *Geophysics*, vol. 69, no. 2, pp. 522–532, 2004.

[3] M. A. Biot, "Theory of Propagation of Elastic Waves in a Fluid-Saturated Porous Solid. I. Low-Frequency Range," *The Journal Of The Acoustical Society of America*, vol. 28, no. 2, pp. 168–178, 1956.

[4] G. Mavko, T. Mukerji, and J. P. Dvorkin, *The Rock Physics Handbook*. 2009.

[5] A. R. Schoepp and G. F. Margrave, "Improving seismic resolution with nonstationary deconvolution 1998 SEG Expanded Abstracts 1998 SEG Expanded Abstracts," in *SEG Annual Meeting*, 1998, vol. September , no. 1.

[6] R. G. Stockwell, "S-transform analysis of gravity wave activity from a small scale network of airglow imagers," The University of Western Ontario, 1999.

Using GIS and Teledetection data to assess mobility and land consumption in polynucleated landscapes

Carlos Marmolejo Duarte

Associate Professor, Centre for Land Policy and Valuations, ETSAB, UPC, Av. Diagonal 649, 08028, Barcelona

Moira Tornés Fernández

PhD candidate, Centre for Land Policy and Valuations, ETSAB, UPC, Av. Diagonal 649, 08028, Barcelona

Corresponding address: carlos.marmolejo@upc.edu

Polycentrism has become the keystone in a major number of regional and urban policies, since it has been seen as a sustainable and equilibrated urban model. In this paper, the problem has been focused in case study the biggest metropolitan areas in Spain and pretends test whether polycentric urban growth does effectively reduce land consumption and travel-to-work journeys, protecting in this way agricultural and forest areas around cities and at the same time reducing energy and air emissions produced by cars. The methodology used has been departed from land-use, transport and census data, and using ArcGIS and TransCAD a group of spatial indicators is calculated and introduced in a family of regression models, where explained variables are per capita land consumption and excess commuting respectively. The results suggest that polynucleation has little effect both in the reduction of land consumption and excess commuting. On the contrary, other variables associated to urban form do highly influence land and mobility patterns, such as fragmentation of urban fabrics, job ratio balance, and the diversity of economic activities and housing offer. Such conclusions may shed light in the design of urban policies, and focus the attention on the definition of small-scale urban variables instead of structural ones at metropolitan scale.

Environmental sustainability, polycentrism, land consumption, excess commuting, Spain

1. Introduction

The spatial arrangement of population and employment is an important issue in contemporary metropolises, particularly in front of the emergence of sprawl and unstructured urban growth, which comprises high risk of environmental damage.

From a social point of view separating employment and residence results into prejudicial consequences for social groups of limited mobility; from an economical point of view distancing firms means losing the opportunity to take advantage of external economies and from an environmental point of view scatteration results in excessive land consumption and environmental costly transport systems. In Spain this latter topic has become especially relevant in the course of the last 15 years, a period in which the real estate sector has produced several times the actual demand for housing, most of the times following a high land consumption scheme. Such a concern has clearly trespassed the national jurisdiction as it is reflected in the Auken Report for European Parliament concerning the impact of extensive urbanisation and environmental menaces particularly in coastal zones (Auken, 2009).

Politicians, especially in Europe, have seen polycentric development as an alternative model to dispersion allegedly leading to cohesion, competitiveness and sustainability (Meijers, 2008); although, the empirical basis of such benefits is still weak and in some cases is contradictory (Boix & Trullén, 2012). In such a way, it has been said that polycentricity has much of normative rather than analytical (Green, 2007). Our main goal is to explore to what extend subcentres do have an impact on neighbouring densities, reducing in this way the land consumed by urban activities, and thus shed light on the environmental sustainability of polycentricity in the regional context of the metropolises analysed. This objective is aligned with Muñiz, et *al.* statement "The importance of polycentrism lies not only in the possibility of concentrating jobs in a limited number of areas under conditions of high density, but also in its capacity to structure and hierarchize urban growth as compared with a dispersed model, amorphous and destructured, without anchorages" (2008; pp. 628).

Although the concept of polycentrism remains highly fuzzy two distinguishable features may be observed: the first is related to the scale of analysis and the second to the conceptualization of subcentres and their influence on their hinterland. In the first, the scales may vary from continental to intrametropolitan scale passing thought a regional scale in between (Kloosterman & Musterd, 2001). In the second subcentres may be analyzed in a morphological or functional way. In this paper the attention is focused in the intrametropolitan scale and necessarily, although not directly, following a morphological approach. From this junction the rise of polycentrism is rather a contrasting that the decentralisation of central cities of the last two centuries has taken a more polycentric *form* at the time that employment subcentres appear on peripheries (Anas *et al.* 1998). In such a way the formerly low density and monofunctional suburbia has become in a more complex post-suburbia having dense multifunctional employment and housing concentrations on the edge of cities (Bontje, 2004).

Using GIS and Teledetection tools we test whether the hypothesis that polycentric urban growth may leads to a more sustainable urban model by reducing the per capita land consumption and journey-to-work trips. For this we use the seven biggest metropolitan areas in Spain. The remaining of the paper is structured as follows: 1) first the theoretic framework on land consumption/density gradients as well as trip generation are presented, 2) after the previous empirical evidence of these issues is reviewed, 3) methodology used in the paper is presented, 4) results are discussed and 5) a general review of the main findings is summarized in the conclusive epigraph.

2. Why polycentrism may impact on land consumption and travel to work journeys?

At the first glance it seems evident that a metropolis that growths by the creation of subcentres is more compact than one that grows by expulsing households and firms to a rapidly expanding sprawl. Nonetheless compacity in a polycentric framework does not only come from the subcentres itself, but mainly from the influence exerted over the neighbouring locations.

The standard urban model as shaped by Alonso (1964), Muth (1969) and Mills (1967) with roots in the pioneering work of Thünen and Laundhardt is the theoretical framework behind the formation of urban densities. This model, originally conceived for a monocentric city, explains that in achieving locational equilibrium households bid for land according to costs saved in commuting. Thus the closer the place of residence to the CBD (where all employment is supposed to be located) the higher the rent transferred to land (which capitalises into higher prices), resulting in a land rent gradient. It is the existence of land rent gradients that underlie the formation of density in a competitive market scenario. If it is considered that house builders invest capital in land and building when developing a site, and constant returns per unit of land are relaxed (i.e. once substitution between land and building cost is allowed), real estate developers economise on the use of land in more central locations where prices peak. In optimising developments they add more building capital per unit of capital invested on land, i.e. they build multi-storey structures instead of low rise ones, resulting in a density gradient following that of land prices. The parallelism between rent and density gradients depends upon the elasticity of substitution between land and capital (Kau & Lee, 1976).

Mills and Hamilton (1984) demonstrated, starting from the monocentric city model, that under certain constraints, such as Cobb-Douglas's production function for housing, users with identical tastes and income and unit price elasticity of demand for housing, density gradients adopt a negative exponential function. According to McDonald's excellent review (1989) it was Stewart (1947) who apparently first empirically used the negative exponential function to test whether population densities decrease with the distance to the CBD; although McDonald notes that it was Clark (1951) who popularised such a function amongst scholars.

In the case of employment density, the negative exponential form is derived by Mills (1969) by assuming that the production functions for product and transportation have also a Cobb-Douglas form, and that the demand for product has constant price elasticity; in this case density decreases as we move further away from the CBD because the access to the agglomeration economies generated there is limited by distance.

On the other hand the urban form (i.e. the spatial distribution of employment and population) should influence commuting patterns. So, in a polycentric framework job places are concentrated in several focal points across the city, thus workers have more opportunities for searching a job place near home. Consequently, *ceteris paribus*, in a polycentric metropolis travel to work journeys should be shorter than those produced in a monocentric system where all workers must commute to the unique employment centre in city.

3. Literature review

3.1. Polycentrism and land consumption

Although the seminal empirical work of Griffith (1981) found not significant effects on population density produced by secondary employment centres in Toronto, it laid the path followed for other scholars in more extended urban areas. Such a path broadly consists in testing whether the influence of subcentres is significant in a polycentric exponential negative density model. Following to Griffith, Gordon & Richardson (1986) found in Los Angeles that for both employment and population the polycentric model fitted better than the monocentric, albeit only 6 of 57 candidate sites were found to influence densities. Although their study related not to employment centres (as bid rent theory suggest) but to residential, received initial criticism (Small & Song, 1994) only recently some authors have recognised that population subcentre may also exert an organising power over the neighbouring density due to the existence of localised amenities: cultural, environmental and consumption amenities (García-López, 2010). Most of the posterior empirical works have found that proximity to employment subcentres do produce an increase of employment densities after dealing with some problems. One of these problems is the presence of "spatial multicollinearity" produced by the introduction of as many distances as subcentres there are (Heikkila, 1988). To avoid such an issue some studies estimate models using only the observations in the area surrounding that centre (e.g. McDonald & Prather; Muñiz, et *al.* 2008); although most of the studies ascribe zones to the nearest subcentre, in any case both approaches relegate the fact that subcentres influence may overlap in interstitial areas. Also assuming that people commute to the nearest subcentre may be risky in the context or metropolitan areas with orographic accidents and non-isotropic transport systems; and in the context of the so called second demographic transition in which households do have more than one person occupied is implausible to assume that all of them commute to the same subcentre (Champion, 2001). On the other hand having only one covariate with the distance to the nearest subcentre makes it impossible to assess the different influence exert by different in size and activity mix subcentres. Some authors (McDonald & Prather, 1994; Small & Song, 1994; McMillen, 2003) create as many variables as subcentres there are and take the inverse of distance to subcentre to minimize multicollinearity assuming that subcentres produce a local influence in front of overall influence exerted by the CBD. McDonald & Prather (*op. cit.)* following such a procedure concluded that distance to O'Hare Airport in Chicago was the second most determinant of population in Chicago after its CBD; also using the same technique Small & Song demonstrated that Los Angeles main centrality is still placed in its downtown and not around the airport as it had previously been stated by Gordon et *al.* (1989).

Influence of subcentres on population density is less conclusive. Some studies have found that subcentres result not significant, and even do exert the inverse effect on neighbouring population density (i.e. density increases with the distance to subcentres). At a first glance, such findings are not completely erratic: if accessibility do account less than negative effects produced by the congested, manufacture-unpleasant or decadent-unsafe centres it is feasible to find attractive dense locations for residence well faraway. Does it mean that the standard theory in urban economics has become out fashioned? Or there are subjacent problems in such analyses? In trying to understanding such findings some explanations have been given: in Chicago McMillen & Lester (2003) found that employment growth was concentrated around pre-existing subcentres outbidding residences to more distant locations; McDonald & McMillen (2000) found also for Chicago that new and big housing developments occurred far away from subcentres due the existence of vacant land in peripheral sites and because subcentres tend to attract commercial developments making land unavailable for housing; Baumont et *al.* (2004) have pointed, in their study for Dijon, the fact that the subcentres are too near to CBD would imply that moving away from them, towards CBD, would mean increasing the density due the more important effect of this later; also having subcentres in the very edge of the city means that they do not exercise any influence in the still not urbanised outward periphery. McMillen (2003) has argued that reversed sign for proximity to subcentres may derives from the use of gross density (all the urbanised land) instead from net density (considering only residential land) which implies that gross residential density is low in centres with most of their land intended for economic activity. For that reasons authors such as Muñiz et *al.* (2003) have used splines to locally adapt the curve of population density to such central depression. McMillen (2003) also has noted that some subcentres may be not yet large enough to influence distribution of population density. For these reasons McMillen & Lester (2003; pp. 78) have concluded that "subcentres are still primarily a non-residential phenomenon"; a conclusion that is clearly contradictory to that delivered by Small & Song (1994) who concluded for Los Angeles that employment subcentres modest influence employment density, but by contrast, population density is strongly influence by them.

3.2. Polycentrism and journey to work

The relationship between the urban form, as studied here and journey to work is less developed. Shwanen *et al.* (2001) have found for the Netherlands that polycentric developments encourage the use of cars since the public transport network is basically intended to radial trips. Nonetheless the proximity of subcentres to CBD produce an influence of journey to work travels as has been demonstrated by Pivo (1993) for the case of Toronto and Aguilera and Mingot (2004) in France, so close-to-CBD subcentres present less car trips, basically due the presence of public transport, namely railroad based. Beyond the presence of public transport system, the very nature of subcentres influence travel to work, Cervero and Wu (1997) found that outlying and low density subcentres favour the use of the car. Nonetheless use of car not necessarily means larger commuters; Cervero and Wu (*Op. Cit.)* found that in low dense and outlier subcentres travels are shorter than travels to large subcentres since the presence of public transport encourages long distance trips.

Shwanen et *al.* (2001) state that the reduction of travel-to-works patterns rely in the nature of labour markets. So if the labour market of subcentres is different to that present in CBD the travel to work distances are reduced, since peripheral centres capture people living around them, conversely, if the labour markets of outlier subcentres and CBD are not complementary but competitive travel to work patters in such polycentric schemes are the same than in monocentric cities.

Guiliano and Small (1993) have discussed whether housing balance (the ratio of employment to population) is determinant in the reduction of commuting patterns. Their empirical findings have pointed out that other parameters such as housing quality and environment are far more important that the commuting distance when people make their location decision. In that respect Wachs *et al*, 1993, claim that the lack of appropriate housing (according to income) in subcentre location may obscure the effect of housing balance in reducing commuting.

4. Case study, data and methodology

4.1. Case studies

In this paper we study the impact of polycentrism on land consumption and labour commuting in the seven biggest cities in Spain: Madrid, Barcelona, Valencia, Bilbao, Seville, Saragossa and Málaga as delimited by Marmolejo et al. (2012). Using travel to work data such authors, using the so called, interaction value, also identify the structure of metropolitan cities, it is to say, the delimitation of main centre and subcentres, and the area structure by them named subsystem. Figures 1 depicts the main figures of the studied cities and figure 2 the structural form. . Barcelona, Valencia and Bilbao stand as the areas with the highest number of subcentres, that conjointly concentrate a significant share of employment (ranging from 20% in Valencia to Barcelona with 22%); at the same time those metropolitan areas do concentrate the lowest share of employment in their expanded-CBD[1] (ranging from 47% in Valencia to 56% in Bilbao). Exactly in the inverse position are Madrid, Seville and Saragossa, which stand as the most monocentric and less polycentric metropolises in Spain. Málaga is an outlier, because having only 4 subcentres they have an important share of employment (23%), and at the same time its expanded-CBD is not to big as in the case of monocentric metropolises.

Figure 1 Main figures of biggest metropolitan areas in Spain

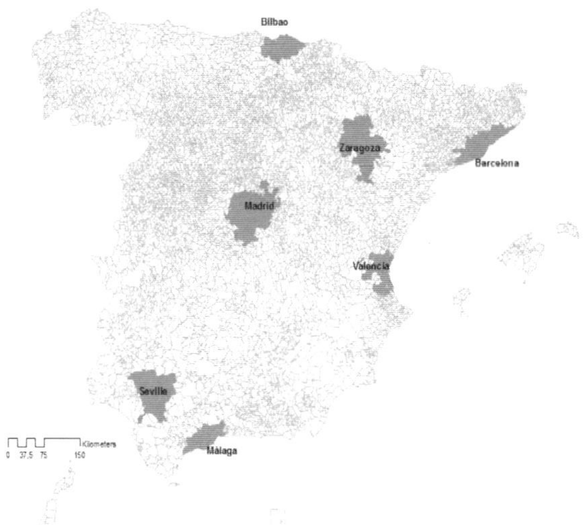

	Number of municipalities	Built up land	Employment	Population	Density
		a	b	c	.=(b+c)/a
Madrid	183	860	2,446,400	5,542,843	9,291
Barcelona	184	745	1,903,867	4,530,164	8,636
Valencia	104	308	686,247	1,792,375	8,046
Sevilla	52	237	447,849	1,381,531	7,719
Bilbao	123	112	445,666	1,231,367	15,024
Zaragoza	88	127	301,860	724,335	8,066
Málaga	32	194	366,525	994,984	7,032

Source: Corine Land Cover & National Census 2001 (ICN, INE)

[1] We compute as extended-CBD those municipalities that: 1) are part of the urban continuous of the main municipality –i.e. their urbanized patches are closer than 200 m- and 2) are inside of the functional subsystem of CBD as detailed above in the main text.

IOP Conf. Series: Earth and Environmental Science **19** (2014) 012006 doi:10.1088/1755-1315/19/1/012006

Figure 2 Structure of the metropolitan areas

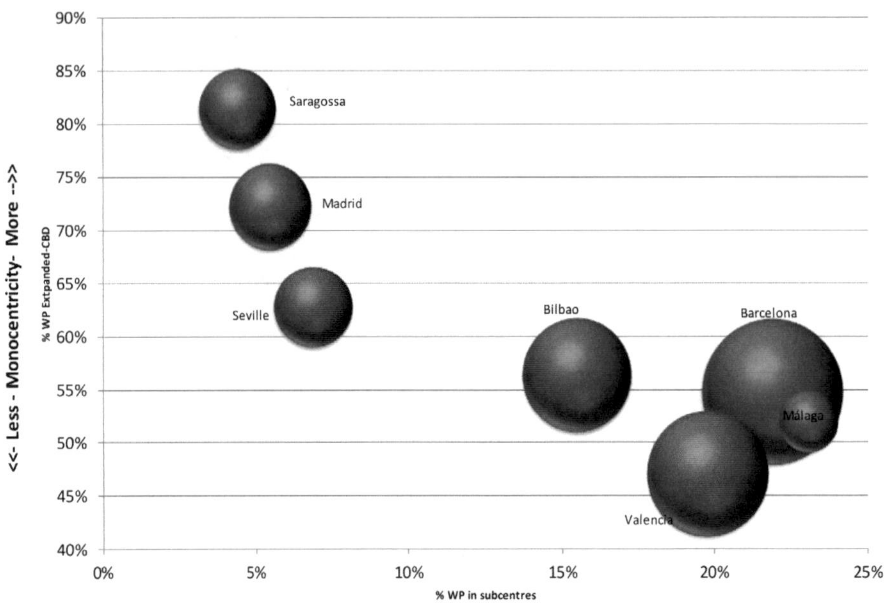

	Employment ('000)	% employment in the expanded-CBD	Number of subcentres	Employment in subcentres (%)	Population in subcentres (%)	Entropy in the employment distribution in all centres	Average number of steeps to integrate all municipalities into the expanded-CBD
Madrid	2,446	72%	8	5%	6%	0.21	1.66
Barcelona	1,904	55%	23	22%	22%	0.83	2.24
Valencia	689	47%	17	20%	19%	0.77	2.07
Seville	448	63%	7	7%	9%	0.25	1.44
Bilbao	438	56%	14	15%	15%	0.50	1.92
Saragossa	302	81%	7	4%	4%	0.14	1.80
Málaga	367	52%	4	23%	16%	0.50	1.66

The size of sphere is significant of the number of subcentres
Source: own elaboration

4.2. Data

We primary use data coming from the National Census 2001 (the last available) at municipal level (the smallest unit for travel to work data at destination). Departing from such a source we use:

1) Travel to work data used to delimit metropolitan areas and identify subcentres as well as characterize commuting patterns
2) Characterize the labour market in terms of industrial classification of sectors, diversity of the economic activities, as well as type of occupation
3) Characterize the income level, departing from the occupation of working population
4) Characterize the housing market departing from the size and quality of houses

Also we use Corine Land Cover to analyse land use patterns. Corine (Coordination of Information on the Environment) Land Cover project for the year 2000, is leaded by the European Environment Agency, and it uses satellite imagery from LandSat and SPOT to photointerpreted the use of land inside the EU. With such information we calculate:

1) The consumed land per capita at municipal level
2) The fragmentation of urban fabrics[2], it is to say the level of discontinuity of the urban tissue

Finally using the Digital Terrain Model we obtain the orography of urban areas. Distances between municipalities are computed using TeleATLAS cartography.

All the information is managed and analysed using ArcGIS (for land use and digital terrain model), TransCAD for travel to work modelling and SPSS for the statistical analysis.

4.3. Methodology

Land consumption

In order to test whether polycentrism influence land consumption we construct a regression model where the dependent variable is population/employment density (the inverse of per capita land consumption) and the explaining variables are distance to CBD and to subcenters. If this latter indicator becomes significant, polycentrism has an influence on land consumption, which is positive in the case that the density gradient is reduced as the distance to the subcenters increase.

Also we introduce some others control variables that have a theoretical impact on land values and consequently on urban densities.

Commuting

In order to prove the relation between polycentrism and commuting patterns we construct the excess commuting indicator, departing from the optimal commuting index of White (1988), that minimizes:

$$CT = \sum_i \sum_j (C_{ij} X_{ij})$$

Where
Cij is the cost of commuting (distance)
Xij is the number of workers that Works in zone *i* and travel to zone *j*

[2] The fragmentation has been computed using the Shannon entropy formula:

$$H_i = -1 * \sum_{x=1}^{n} Px_i * \ln(Px_i)$$

In this case *P* is the probability to find urbanized land in a given *x* spot in a x municipality. In a x municipality are as many spots as urban patches are. If two patches are separated by a gap inferior to 200m it is considered that form part of the same patch. This later criteria allows for consider the interruptions caused by rivers and other lineal infrastructures (e.g. high voltage electric lines).

Put in simple, such excess commuting index compares the optimal commuting (all the workers commute to the nearest available job place) to the actual commuting. The bigger the indicator is, the higher the unnecessary commuting is. The calculus of optimal commuting has been performed in TransCAD[3] software using the built-in optimization model.

Finally, we do a regression model of excess commuting indicator over indicators of urban form: housing balance, distance to CBD and subcentre, transport facilities, income level, employment mismatching (the level of coherence between the employment and work force qualification), employment diversity.

5. Results

5.1. Land consumption

Until know the analyses have depicted an image of divergent metropolises, is time now to study what extend polycentric urban growth is more sustainable from the perspective of land consumption. The average land consumption (built up area/(employees + residents)) in CBDs is 52 sq. m. per capita, while in subcentres it is almost three times bigger equivalent to 143 sq. meters per capita and finally in the remaining of zones it reaches 178 sq. meter per each resident and employee. Whereas such findings are not surprising at all, it remains unclear whether subcentres exert any influence on land consumption beyond themselves. Table 1 reports the result of polycentric semi-log models in which the dependent variable is the logarithm of the urban density (i.e. the inverse of land consumption per capita) and the independent variables are those related to accessibility. In order to avoid multicollinearity problems regarding to the subcentres lying near the CBD, the inverse of the distance to the functional subcentre is taken. In doing that it is assumed that subcentres exert a more localised influence on urban densities in front of the overall influence of CBD. One differentiating point in relation to previous studies is the adscription of each zone to the functional subcentre and not to the nearest. To avoid endogenous problems both the CBD and the subcentres are not considered in the analysis, such elimination also corrects the bias that would introduce the inclusion of central municipalities with a bigger area due their administrative roles; also this exclusion is justified since what is pursued here is the impact of proximity to centres.

Using only accessibility variables it is possible to explain urban densities *only* in 5 of the 7 studied cities, albeit the models for Bilbao and Saragossa poorly explain the inverse of per capita land consumption. Proximity to CBD exert a very similar influence in the biggest cities, both in Madrid, Barcelona and Valencia urban density declines about 3% for each km that the distance to their CBDs increases. Nevertheless in Bilbao and Saragossa proximity to the CBC seems not to exert any

[3] The sources of information are two, in terms of demographics and mobility from residence to work in 2001 are extracted from the National Statistics Institute (INE) of Population and Housing provided by the Census and in terms of infrastructure the road network of Tele Atlas year 2001. With all of these three matrices of travel flows have been developed with TransCAD 5.0 software. In this software has been worked with three covers, one with municipal information, another with the network calculated with Tele Atlas and other last, that of centroids of each municipality, which represents the centre of gravity of each municipality.
With data and cover three matrices of travel flows are calculated. First, the current commuting between each centroid of each municipality. The second, the distance matrix, in kilometres, where distances that cross population of a municipality of residence to another to work are shown. And finally the third matrix, the optimal commuting matrix, that means, repositions in simulated way so people have to travel the shortest distance (minimal cost) to get from his home to his work.

influence over urban density. The null or inverse influence of CBD on urban densities is not rare, McMillen (2001) found in Dallas, Houston and San Francisco that the CBD gradient was positive, suggesting that their CBD is no longer the critical determinant of the broad spatial trend in densities, and that a more realistic specification would be treat the CBD as a simply another of the multiple centres in these metropolitan areas.

On the other hand, proximity to subcentres exerts different influences. In Barcelona and Madrid the influence of subcentres is very similar, although the latter denotes a bigger coefficient, such divergence might come from the fact that Madrid has only 8 subcentres and Barcelona 24 distributed in a very similar area, so it is possible that in Barcelona the influence of subcentres overlaps due the proximity between them. The influence of subcentres on urban density in Bilbao and Saragossa is bigger than in the remaining of the cities, what is important to note is that in absence of the influence coming from CBD in these two cities subcentres appear as clear structuring elements of the urban form, although in general poorly explain the overall metropolitan density. Whether such a cities have achieve a more subtle level of morphological polycentrism remains on the result of these findings.

Table 1 Models for urban density using only proximity to functional subcentres

	Barcelona			Bilbao			Madrid			Málaga			Seville			Valencia			Saragossa		
r2 adj	0.38			0.06			0.28									0.26			0.09		
F ANOVA	57.42			6.37			33.36									30.94			8.39		
	B	Beta	t-value	B	Beta	t-value	B	Beta	t-value	B	Beta	t-value	B	Beta	t-value	B	Beta	t-value	B	Beta	t-value
Constant	9.26		63.82	8.83		37.88	8.89		46.32							9.46		52.88	7.40		56.06
Dist CBD	-0.03	-0.60	-9.79				-0.03	-0.51	-7.86							-0.03	-0.52	-5.56			
Inv Dist Sub	3.68	0.22	3.63	4.90	0.27	2.52	3.82	0.16	2.40										6.11	0.31	2.90

Only variables significant at 95% of confidence are presented

As it has been said, proximity to centres appears as an incomprehensive factor to explain urban densities for a number of reasons. For example low-density housing (e.g. garden city fashion) are placed general in the outskirts of the cities, but when they growth and integrate subcentres, as it is quite common in Mediterranean metropolises (in our sample 65% of the subcentres can be considered as integrated), such a low density developments get located in central locations or in between centres. Such an overlapping process related to the arrangement and typologies of urban development may obscure the influence of proximity to centres: for example it is very well know that in Málaga area the low density developments stand out as the primary model of urban growth with independence of their position in relation to the centres. In the cities analysed the correlation between the index measuring the presence of low-density-fragmented developments and distance to CBD is r=-0.01 and r=-0,051 to subcentres but is not statistical significant in both cases.

Also it becomes unclear in such large cities whether distance to centres is masking other variables such as those regarding the topographic nature of territories, as bigger the pressure to urbanise land, higher the probability to use far-from-the-CBD-less suitable zones like hills and mountains in which high-rise development is difficult/costly. In the cities analysed the correlation between distance to CBD and orographic complexity is positive r=0,248 (sig=0.000) and the correlation between distance to CBD and slope is also positive r=0,146 (sig=0.000). What confirms that ancient urban settlements (which eventually became in centres) where located in the plateaus of valleys or along hydric basins. So increasing the distance to such centres means also increase the difficulty/cost to urbanise with high densities, and not only proximity (transport cost saving) to centres.

Table 2 reports the results of models that take into consideration, besides proximity to centres, other control variables.

Table 2 Models for urban density considering proximity to functional subcentres and other control variables

	Barcelona			Bilbao			Madrid			Málaga			Seville			Valencia			Saragossa		
r2 adj	0.59			0.06			0.44			0.59			0.36			0.66			0.09		
F ANOVA	33.26			6.37			34.82			13.21			9.00			28.50			8.39		
	B	Beta	t-value	B	Beta	t-value	B	Beta	t-value	B	Beta	t-value	B	Beta	t-value	B	Beta	t-value	B	Beta	t-value
Constant	9.22		36.10	8.83		37.88	9.60		44.07	13.37		6.92	9.21		64.18	8.94		74.85	7.40		56.06
Dist CBD	-0.03	-0.46	-6.47				-0.03	-0.60	-8.74												
Inv Dist Sub	1.87	0.11	2.08	4.90	0.27	2.52													6.11	0.31	2.90
Suburban railway							0.47	0.20	3.14							0.32	0.18	2.30			
Orography complexity										-2.09	-0.29	-2.17	-0.34	-0.31	-2.47	-0.43	-0.37	-4.53			
Slope	-2.83	-0.19	-3.15																		
Coast	0.46	0.17	2.94																		
Fragmentation	-0.31	-0.17	-2.97				-0.62	-0.33	-5.39	-0.97	-0.72	-5.29	-0.53	-0.52	-4.16	-0.61	-0.40	-5.58			
Tertiary/non-manuf.	0.26	0.20	2.79													0.26	0.23	2.84			
Medhigh-high income																-0.28	-0.25	-3.32			
Medlow_low income	0.22	0.18	3.01				0.22	0.28	4.79	0.63	0.62	4.65	0.16	0.26	2.07	0.25	0.23	3.38			

Only variables significant at 95% of confidence are presented

Unlike the previous models, in all cities it is possible to explain urban densities: in all of them, but in Bilbao and Saragossa, the percentage of the variance explained increases. The sing of the other control variables are as expected when they are present in the models: the dummy controlling the existence of suburban rail stations is positive indicating that municipalities that have such a transport system are more dense; the dummy indicating that a municipality is on the coast is positive indicating the historical effect produced by sea accessibility in the past and the sea externalities in the present; the orography complexity and slope indicators depicts a negative sign which implies that in complex territories the density is reduced; the index of fragmentation of urban fabrics that is a proxy for urban policies allowing low dense/sprawled developments appears with the expected negative sign; the factor synthetizing the presence of service sectors/absence of manufacturing is positive indicating that in tertiary areas the per capita land consumption is reduced due such activities consume less urban land in relation to manufacturing; and finally the income factors are coherent with the hypothesis that wealthy households prefer big houses in detached styles reducing urban densities and conversely poor households only can afford small dwelling most of them in apartment-like arrangements.[4]

The effect of proximity to subcentres in urban density falls in the two biggest metropolitan systems. In Barcelona other variables such as the structure of economic activity, the fragmentation of urban fabric, and those related to the orographic complexity/slope of territories become more important that proximity to subcentres. In Madrid the proximity to subcentres completely fadeout in front of other variables such as: urban fragmentation, the income of population or the presence of suburban railway stations (linking suburban zones basically with the CBD due the radial structure of the network). It is important to note that in both cities proximity to CBD remains as the principal factor explaining urban densities according beta coefficients. Both the permanence in Barcelona of the inverse of the distance to subcentres (which was not significant in the Madrid Model), as well as the slightly smaller beta coefficient for proximity to CBD in relation to Madrid (-0,46 for Barcelona and -0,60 for Madrid), support the idea laid before that Barcelona exhibit a higher level of polycentrism, not only because it

[4] Please note that the inclusion of medium high-high income variable is not incompatible with the introduction of medium low-low income variable since they are orthogonal due they are obtaining by means of a principal component analysis.

has more subcentres in relation to Madrid, but also because its subcentres exert a stronger influence over their functional subsystems, and its CBD a lower influence, once all the other control variables are taken into consideration. *In Barcelona both the daily life of commuters (which produce subcentres in our methodological approach) as well as urban form is more dependent of subcentres in comparison to Madrid where its CBD still exert an extraordinary influence on such variables.*

Only in Bilbao and Saragossa proximity to subcentres retain its explicative power since the remaining of the control variables fail to enter to the model; albeit, as said before, the overall explicative power is very poor. One could argue that distance to centre (subcentre) reduce its importance when topography measures are considered since distance do not include the fact that streets in hilly terrains are longer because they have to adapt the pendent to the technical possibilities of vehicles; nonetheless in our distance measure such irregularities are already taken into consideration since TransCAD measures actual paths following streets. Although such precaution has been taken, in order to prove that *effectively* topography has incidence on urban densities, an alternative set of models have been constructed taken into consideration the *time* to centres calculated in TransCAD, and consequently to include the fact that vehicles slowdown they velocity in sloped longer streets. The result of such a set of models suggest that, as well as in models reported in table 2, distance to subcentres reduces its importance, and even it tends to disappear in Barcelona and Madrid. *These results suggest that terrain topography adds new information to the explanation of urban densities.*

Table 3 Models for employment and population density considering proximity to functional subcentres and other control variables

Employment density

	Barcelona			Bilbao			Madrid			Málaga			Seville			Valencia			Saragossa		
r2 adj	0.57			0.17			0.53			0.47			0.32			0.65			0.21		
F ANOVA	34.88			9.24			38.88			12.14			7.80			31.64			8.04		
	B	Beta	t-value	B	Beta	t-value	B	Beta	t-value	B	Beta	t-value	B	Beta	t-value	B	Beta	t-value	B	Beta	t-value
Constant	7.85		27.49	7.55		30.33	8.14		30.93	7.37		56.26	7.75		31.56	7.41		53.90	6.35		23.40
Dist CBD	**-0.03**	**-0.52**	**-7.50**				-0.04	-0.66	-10.52										-0.01	-0.25	-2.30
Inv Dist Sub	2.66	0.15	2.64	5.53	0.29	2.80	**5.09**	**0.17**	**3.03**										5.72	0.22	2.11
Suburban railway							0.54	0.19	3.20							0.58	0.29	3.83			
Orography complexity																**-0.67**	**-0.50**	**-6.24**			
Slope	**-3.61**	**-0.22**	**-3.62**										-8.67	-0.37	-2.81						
Coast	0.50	0.17	2.85	-0.86	-0.31	-3.07															
Fragmentation							-0.49	-0.21	-3.80	-0.62	-0.56	-3.72	-0.38	-0.33	-2.54	-0.46	-0.27	-3.61	0.24	0.24	2.32
Tertiary/non-manuf.	0.23	0.16	2.18																		
Medhigh-high income																-0.20	-0.16	-2.17			
Medlow_low income	0.26	0.20	3.16				0.22	0.24	4.07	0.51	0.61	4.04	0.21	0.30	2.25	0.30	0.24	3.50			

Population density

	Barcelona			Bilbao			Madrid			Málaga			Seville			Valencia			Saragossa		
r2 adj																					
F ANOVA																					
	B	Beta	t-value	B	Beta	t-value	B	Beta	t-value	B	Beta	t-value	B	Beta	t-value	B	Beta	t-value	B	Beta	t-value
Constant	9.15		43.97	8.55		35.15	9.23		42.19	14.03		6.93	8.97		62.27	8.78		80.07	7.02		49.49
Dist CBD	**-0.02**	**-0.43**	**-5.95**				-0.03	-0.56	-7.91												
Inv Dist Sub				4.31	0.23	2.13													5.00	0.25	2.34
Suburban railway							0.46	0.20	3.05												
Orography complexity										**-2.51**	**-0.32**	**-2.49**	-0.38	-0.35	-2.80	-0.40	-0.34	-3.89			
Slope	-2.99	-0.20	-3.36																		
Coast	0.48	0.18	3.01																		
Fragmentation	-0.35	-0.19	-3.31				-0.66	-0.36	-5.68	-1.04	-0.73	-5.46	-0.51	-0.50	-4.00	-0.60	-0.38	-5.04			
Tertiary/non-manuf.	**0.28**	**0.22**	**2.91**													0.38	0.33	3.94			
Medhigh-high income																-0.26	-0.22	-2.79	-0.30	-0.30	-2.88
Medlow_low income	0.20	0.17	2.64				0.21	0.27	4.47	0.66	0.61	4.67				0.21	0.19	2.54			

Only variables significant at 95% of confidence are presented
In bold are non stationary variables (at least a 90% of confindence) according to the Test of Montercarlo in the LWR analysis

The remaining of the coefficient variations are as expected, but some explanations are needed, for example: the slope variable seems to reduce more the employment density since steeper areas are less suitable for economic activities (mostly for those extensive); urban fragmentation is more correlated with low-density housing fabrics since those urban-sprawl-developments are primarily residential, while the factor synthetizing the presence of tertiary activities seems to positively affect more the population density than economic activity. This latter paradox is solved when it is considered that in compact cities (like these analysed) service sectors and housing coexist in the same buildings (because externalities of services are quite compatible with residential uses) in dense fabrics, for instance using the ground level for retail, the first levels for office activities and the remaining for flats. So increasing the proportion of service activities also means an increase in the density of population. The same is valid for medium low-low income (with a higher positive incidence on the employment density), since mixed activity buildings are not seen as exclusive high standing alternatives for living, and rarely are chosen by high-income households, who prefer only-housing-buildings and even only-housing-neighbourhoods

5.2. Commuting

Regarding the relationship between commuting and polycentrism the inferior table summarises the results of a family of models where the explained variable is the natural log of excess commuting. The first model "Transport" is able to explain only 3% of the variance of excess commuting, according to such a model the higher is the presence of railway stations (most of them rendering a suburban-radial train service), the higher the excess commuting, such a finding suggest that working population living in well-connected areas serviced by high capacity transport network do travel more that those living in poorly connected areas. The second model "urban form" is able to explain 19% of the excess commuting, which is relevant to the interest of this research since its explanatory capacity is the highest among individual models. According to such a model, the higher is the presence of manufacturing activity the higher is the excess commuting; such a finding is relevant, since during the last four decades in Spain, as well in other parts of the world, most of the new and decentralising economic activity has been accommodated in industrial parks located in suburban places. The positive sign of the coefficient suggest that manufacturing plants does not encourage the self-contention of site's working population, on the contrary those municipalities depicting a high level of such activities denote the highest commuting patterns, and behind this issue is the fact that manufacturing locations are well serviced by motorways connecting them with the remaining of the metropolitan system. The second coefficient is the dwelling diversity, such an indicator represents the diversity of housing in terms of size (as a proxy of housing typologies), the negative relationship with excess commuting suggest that well developed residential areas (with a diverse offer of dwellings matching different income levels) do have a higher self-containment of commuters, since they are able to find the house they can afford or that fulfils their residential expectative. The mismatching CNO coefficient is significant of the non-correspondence between the working population and the employment in a given site in occupational terms. The higher this coefficient is, the bigger the mismatch between the offer and demand of jobs in qualification terms is. The positive correlation of this index and excess commuting ratifies that very specialized job places, which labour force do not match the qualification required by firms do produce higher commuting patterns that balanced zones. Finally the job ratio (the number of job places to working population) confirms that very economic specialized zones (e.g. manufacturing parks); paradoxically do not contribute to the reduction of commuting, the reversed sign of the square of job ratio suggest and exponential function in the relation with excess commuting.

The income model is constructed over the principal component analysis that summarizes the socio-professional structure of working population. In such a factorial analysis high income is assumed to be related with managerial and professional working population, medium income is related to workers employed in the personal service sectors, and medium-low profiles includes also the medium qualified manufacturing working population. As suggested by the model there is not a linear relationship with excess commuting and income as theory suggest, high qualified workers depict a negative relationship with excess commuting which indicates that these professional profiles tend to live near their job locations, exactly the same is true for medium-low profiles. The operational principia is the same in both cases, since their income make wealthiest workers live in expensive locations near office based jobs, at the time that blue-collar-workers only can afford housing in the poorest residential areas that are quite often located next to manufacturing locations. On the contrary, medium income population seems to have longer commuting patterns, since their employment oriented to people services is more sprawled across city.

In the integrated model all the precedent models are integrated in a unique one. Some of the variables are introduced and other eliminated. For example in the transport infrastructure dimension the motorway service (expressed as the number of entries by 10.000 inhabitants) is introduced with the expected positive sign, in the urban form dimension distance to CBD is introduced with a positive sign which suggest that peripheral municipalities show higher excess commuting as expected, other control variables include the orographic complexity with a positive sign that reveals that hilly territories increase the commuting patterns and dummies controlling the cities.

As observed the distance to subcentre does not have any influence on excess commuting, as well the dummy representing the subcentre is not introduced in any model.

Table 4 Excess commuting models

		MOD1		MOD2		MOD3		MOD4	
		Transport		Urban form		Income		Integrated	
	R	0,17		0,44		0,24		0,52	
	Sq R	3,0%		19,2%		5,9%		27,3%	
	Sq R adjusted	2,9%		18,7%		5,5%		26,2%	
	Error típ.	0,94		0,86		0,93		0,82	
		B	t	B	t	B	t	B	t
Dim	Constant	1,06	29,82	1,90	5,07	1,10	31,85	0,86	2,23
T	Stations/10.000 inha	0,03	4,74						
	Motorway acc/10.000 inhab							0,00	3,49
UF	% manufacturing			1,88	8,95			1,04	4,65
	Dwelling diversity			- 0,90	- 4,77			- 0,49	- 2,51
	Mismatching C N O			0,57	3,44				
	Job ratio			- 0,02	- 4,14				
	Job ratio^2			0,26	3,62			- 0,03	- 5,80
	LTL/Viv tot							0,51	5,66
	Distance to CBD							0,01	3,23
TM	Orographic complexity							0,14	2,34
INC	Medium low					-,169	-4,853		
	Medium					,119	3,380		
	High					-,112	-3,239		
I	Bilbao							0,41	3,83
	Valencia							0,59	5,86
	Málaga							- 0,47	- 2,86
	Zaragoza							0,34	3,32

T= Transport, UF= Urban Form, TM= Territorial matrix, INC=income, I=dummy for diferent cities

Conclusions

Concentrated decentralization of employment and housing in a polycentric scheme has been seen as a sustainable alternative to monocentrism where all employment and services remain concentrated in a unique location obligating people to commute for working and other purposes. Polycentrism has been also seen as an alternative to scatteration where both employment and housing sprawl across the city in low dense scheme making it impossible to provide sustainable transport systems and consuming high quantities of land per capita. Nonetheless such a posture is more normative than analytical. Using GIS and Teledetection technologies in this paper we investigate whether polycentric urban growth does contribute to reduce land consumption and excess journey-to-work commuting. The problem has been answered through our empirical programme carried out for the seven biggest metropolitan areas in Spain gives little support to the role of polycentrism in the reduction of the aforementioned issues.

- In reducing land consumption other urban features, such as the typology of urban fabrics, have a higher impact rather than the number or proximity to subcenters.

- As well, in commuting terms the diversity of housing estates and diversity of employment encourage the reduction of excess commuting.

Such findings suggest that policy makers should change the target of their normative recommendations, putting more attention to other aspects of urban form rather than the way in how employment is spatially distributed across the city.

References

[1] Aguilera, A.; Mignot, D. (2004) Growth in Commuting Distances in French Polycentric Metropolitan Areas: Paris, Lyon and Marseille Urban Studies **42**, pp. 1537-1547.

[2] Alonso, W. (1964). Location and Land Use, Cambridge, Mass. Harvard University Press.

[3] Anas, A.; Arnott, R.; Small, K. (1998). Journal of Economic Literature **36**, No. 3, pp. 1426-1464.

[4] Auken, M. (2009) Report: On the impact of extensive urbanisation in Spain on individual rights of European citizens, on the environment and on the application of EU law, based upon petitions received, European Parliament.

[5] Beaumont, C.; Ertur, C.; Le Gallo, J. (2004) Spatial convergence clubs and the European growth process, 1980-1995», in B. Fingleton (ed): European regional growth, Springer Berlin, Heidelberg New York, pp. 131-58.

[6] Boix, R.; Trullén, J. (2012). Polycentrism and urban structure: a critic review from the perspective of the research agenda, ACE: Architecture, City and Environment **6**, 18.

[7] Bontje, M. (2004) From suburbia to post-suburbia in the Netherlands: potentials and threats for sustainable regional development. Journal of Housing and Built Environment **19** (1): pp. 25-47.

[8] Cervero R.; K.L. Wu (1997) Polycentrism, commuting and residential location in the San Francisco Bay area. Environment and Planning A **29**, pp. 865-886.

[9] Champion, A.G. (2001). A changing demographic regime and evolving polycentric urban regions: consequences for the size, composition and distribution of city population. Urban Studies **38**, No. 4, pp. 657-667.

[10] Clark, C. (1951) Urban population densities. Journal of the Royal Statistical Society: Series A 114, pp. 490-496.

[11] Garcia-Lopez, M.A. (2010) Population suburbanization in Barcelona, 1991–2005: Is its spatial structure changing? Journal of Housing Economics **19**, pp. 119-132.

[12] Giuliano G., Small K.A. (1993) Is the Journey to Work Explained by Urban Structure? Urban Studies **30**, No. 9, pp. 1485-1500.

[13] Gordon, H; Ricardson, J., Giuiliano, G. (1989) Travel Trends in non-CBD Activity Centres, Final Report, Urban Mass Transportation Administration, U.S. Departament of Transportation.

[14] Gordon, P., Richardson, H.W.; Wong, H.L. (1986) The distribution of population and employment in a polycentric city: the Case of Los Angeles, Environment and Planning A **18**, pp. 161-173.

[15] Green, N. (2007) Functional Polycentricity: A Formal Definition in Terms of Social Network Analysis, Urban Studies **44**, No. 11, pp. 2077-2103.

[16] Griffith, D.A. (1981) Modelling Urban Population Density in a Multi-centered City, Journal of Urban Economics **9**, pp. 298-310.

[17] Heikkila, E. (1988) Multicollinearity In Regression Models with Multiple Distance Measures, Journal of Regional Science **28**, No. 3, pp. 345-362.

[18] Kau, J. B.; Lee, C. F. (1976) Capital-land Substitution and urban land use. Journal of Regional Science **16** (1): pp. 83-92.

[19] Kloosterman, R.; Musterd, S. (2001) The Polycentric Urban Region: Towards a Research Agenda, Urban Studies **38**, No. 4, pp. 623-633.

[20] Marmolejo, C.; Masip, J.; Aguirre, C. (2012) Policentrismo en el sistema urbano español: un análisis para 7 áreas metropolitanas. En: Ciudad y Territorio, Estudios Territoriales.

[21] McDonald, J. F., (1989) "Econometric studies of urban population density: A survey", Journal of Urban Economics **26**, pp. 361-385.

[22] McDonald, J. F. & McMillen, D. P., (2000) Employment Subcenters and Subsequent Real Estate Development in Suburban Chicago. Journal of Urban Economics **48** (1), pp. 135-157.

[23] McDonald, J.; Prather, P. (1994) Suburban employment centres: The case of Chicago, Urban Studies **31**, pp. 201-218.

[24] McMillen, D. (2001) Non-Parametric Employment Subcenter Identification, Journal of Urban Economics **50**, pp. 448-473.

McMillen, D. (2003) Employment subcenters in Chicago: Past, present, and future, Federal Reserve Bank of Chicago, Economic Perspectives **2**, pp. 1-13.

[25] McMillen, D.; Lester, W. (2003) Evolving subcenters: employment and population densities in Chicago, 1970–2020, Journal of Housing Economics **12**, pp.60-81.

[26] Meijers, Evert (2008) Measuring Polycentricism and its Promises, European Planning Studies **16**:9, pp. 1313-1323.

[27] Mills E.S.; Hamilton B.W. (1984) Urban Economics, Scott Foresman, Glenview, IL.

[28] Mills, E. (1967) An Aggregative Model of Resource Allocation in a Metropolitan Area. American Economic Review **LVII**, No. 2, pp. 197-210.

Mills, E. (1969) The value of urban land in The Quality of the Urban Environment (H. Perloff, ed.) Johns Hopkins Press, Baltimore, pp. 231-253.

[29] Muñiz, I.; Galindo, A.; García-López, M.A. (2003) Cubic Spline Density Functions and Satellite City Delimitation: The Case of Barcelona, Urban Studies **40**, pp. 1303-1321.

[30] Muñiz, I; García-López, M.A.; Galindo, A. (2008) The Effect of Employment Subcentres on Population Density in Barcelona Urban Studies 45, No. 3, pp. 627-649.

[31] Muth, R. (1969) Cities and Housing. Chicago: Chicago University Press.

[32] Pivo G., (1993) A Taxonomy of Suburban Office Clusters: The Case of Toronto, Urban Studies **30**, No. 1, pp. 31-49.

[33] Schwanen, T., Dijst, M. and Dieleman, F.M. (2001) Leisure Trips of Senior Citizens: Determinants of Modal Choice. Tijdschrift voor Economische en Sociale Geografie **92**, pp. 347-360.

[34] Small, K.; Song, S. (1994) Population and Employment Densities: Structure and Change, Journal of Urban Economics **36**, 292-313.

[35] Stewart, J. (1947) Suggested Principles of Social Physics. Science **106**, pp. 179-80.

[36] Wachs M. Taylor B.D., Levine N., Ong P. (1993) The Changing Commute: A Case-study of the Jobs-Housing Relationship over Time, Urban Studies **30**, No. 10, pp.1711-1729.

[37] White, M. (1988) Location Choice and Commuting Behavior in Cities with decentralized employment, Journal of Urban Economics **24** pp. 129-152.

Determination of karst collapse intensity indicator in area of nuclear power plant construction using incomplete data

R Sharapov

Vladimir State University, Orlovskaya str., 23, Murom, Russia

E-mail: info@vanta.ru

Abstract. The paper deals with the definition of karst collapse intensity. The technique for determining the intensity of karst formation and collapse on the basis of calculation and probabilistic method is given. Karst collapse formation is affected by a great variety of natural and anthropogenic factors. Each factor can vary quite widely. The paper describes a technique for determining karst collapse intensity from incomplete data. It uses karst processes monitoring data in the area and monitoring data of areas with similar values of the most significant factors leading to the karst collapses. The method used for determination of karst collapse intensity indicator in area of Nizhny Novgorod nuclear power plant construction.

1. Introduction

Karst-suffusion processes are treated to a number of dangerous geological processes of natural or technogenic origin. They occupy large territories. Exclude such territories from the development and use is not possible. Therefore actual problems are karst processes observing and the assessment of karst hazard of territory and reducing to a minimum the risk of accidents on the newly constructed buildings and structures, caused by the development of karst processes [1, 2].

In observing the karst processes actively used various indicators studied manifestations of exogenous processes. One is karst collapse intensity indicator. It belongs to a group of quantitative indicators of karst processes [3].

Karst collapse intensity indicator describes the average annual number of karst forms on the 1 km^2. It is measured in units/m^2·year [1]:

$$\lambda = \frac{n}{S \cdot t},$$

where n – number of karst forms, S – space of area (m^2), t – the number of observation years (year).

Karst collapse intensity indicator can be easily determined for the areas which are carried out systematic observations of karst processes. Unfortunately, in actual practice often we have to deal with incomplete data obtained during periodic observations of small areas of the territories. Sufficiently accurate information on karst collapses there are for well-established areas. For underdeveloped areas such information may be missing. Thus, there are some difficulties with the definition of karst collapse intensity indicator to territories with slightly studied areas.

Purpose of paper is determinate of karst collapse intensity indicator using incomplete data.

2. Method for determination of karst collapse intensity indicator using incomplete data

The formation of karst collapses affects a large variety of natural and anthropogenic factors. Despite the fact that each area is exposed to a set of factors, it is possible to find small areas, characterized by the same values of the major influencing factors. This can be used to assess the karst hazard of area. In [4] proposed a method for determination of karst collapse intensity indicator, depending on the combination of the values of natural factors. It is shown that the distribution of mutually independent karst collapses in a territory for a certain period of time subject to the Poisson distribution. Karst collapse intensity indicator corresponds to the expected value of karst collapses number in the study area for some time.

Suppose that the flow of karst processes affect n factors weakly dependent on each other. Then:

$$\lambda = \sigma^2 = \sigma_1^2 + \sigma_2^2 + \ldots + \sigma_n^2 + \sigma_{rem}^2, \tag{1}$$

where σ^2 – the variance of karst collapse intensity indicator, σ_1^2, σ_2^2, …, σ_n^2 – variances, caused by the n factors, σ_{rem}^2 – the remain variance.

In the case study areas of the territory with the same sign of some factor we can form the expressions:

$$\begin{gathered}
\lambda_{11} = \sigma_{11}^2 = \sigma_{2-11}^2 + \sigma_{3-11}^2 + \ldots + \sigma_{n-11}^2 + \sigma_{rem-11}^2 \\
\lambda_{12} = \sigma_{12}^2 = \sigma_{2-12}^2 + \sigma_{3-12}^2 + \ldots + \sigma_{n-12}^2 + \sigma_{rem-12}^2 \\
\ldots \\
\lambda_{ij} = \sigma_{ij}^2 = \sigma_{1-ij}^2 + \sigma_{2-ij}^2 + \ldots + \sigma_{n-ij}^2 + \sigma_{rem-ij}^2 \\
\lambda_{nk} = \sigma_{nk}^2 = \sigma_{1-nk}^2 + \sigma_{2-nk}^2 + \ldots + \sigma_{n-1-nk}^2 + \sigma_{rem-nk}^2
\end{gathered} \tag{2}$$

where λ_{11} – expected value of karst collapse intensity indicator on the territory with sign 1 of the factor 1, σ_{11}^2 – variance of karst collapse intensity indicator on the territory with sign 1 of the factor 1, λ_{ij} – expected value of karst collapse intensity indicator on the territory with sign j of the factor *i*, σ_{ij}^2 – variance of karst collapse intensity indicator on the territory with sign *j* of the factor *i*, σ_{1-ij}^2 – variance caused by the variability signs of factor 1 in the area with sign *j* of the factor *i*, σ_{rem-ij}^2 – remain variance in the area with sign *j* of the factor *i*.

The values of the residual variance can be represented by expressions:

$$\begin{gathered}
\sigma_{rem-11}^2 = \sum_{h=1}^{n} \bar{\sigma}_{h-11}^2 + \sigma_{oth.11}^2, \\
\sigma_{rem-12}^2 = \sum_{h=1}^{n} \bar{\sigma}_{h-12}^2 + \sigma_{oth.12}^2, \\
\ldots \\
\sigma_{rem-ij}^2 = \sum_{h=1}^{n} \bar{\sigma}_{h-ij}^2 + \sigma_{oth.ij}^2,
\end{gathered} \tag{3}$$

where $\bar{\sigma}_{h-ij}^2$ – the average variance value that occurs due to variability signs values of factor *h* inside these sign *j* of the factor *i*, $\sigma_{oth.ij}^2$ – variance caused by other unaccounted factors.

Various factors have different effects on the formation of karst collapses. Therefore, we can choose *n* most significant factors. Effect of other factors can be shared equally among the first terms of (3). Then the expression takes the form:

$$\sigma^2_{oth-ij} = \sum_{h=1}^{n} \overline{\sigma^*}^2_{h-ij} \tag{4}$$

Suppose that A_{ij} – the area of distribution of sign j of the factor i, A_{hg-ij} – area of the sharing distribution of sign g of factor h and the sign j of factor i. Then

$$\sum_{g=1}^{k_h} A_{hg-ij} = A_{ij} ,$$

where k_h – signs count of factor h.

The average variance has the form:

$$\overline{\sigma}^2_{hg-ij} = \frac{1}{A^2_{ij}} \sum_{j=1}^{k_i} \sigma^2_{ij} A^2_{hg-ij} \tag{5}$$

Suppose that $A_{hg-ij} / A_{ij} = \alpha_{hg-ij}$, then equation (5) has the form:

$$\overline{\sigma}^2_{hg-ij} = \sum_{j=1}^{k_i} \sigma^2_{ij} \alpha^2_{hg-ij} \tag{6}$$

Then equation (3) can be represented as:

$$\begin{aligned} \sigma^2_{rem-11} &= \sigma^2_{11} + 0 \cdot \sigma^2_{12} + \ldots + \alpha^2_{11-nk} \cdot \sigma^2_{1h} \\ \sigma^2_{rem-12} &= 0 \cdot \sigma^2_{11} + \sigma^2_{12} + \ldots + \alpha^2_{12-nk} \cdot \sigma^2_{1h} \\ &\ldots \\ \sigma^2_{rem-nk} &= \alpha^2_{nk-11} \cdot \sigma^2_{11} + \alpha^2_{nk-12} \cdot \sigma^2_{12} + \ldots + \sigma^2_{1k} \end{aligned} \tag{7}$$

The number of equations is equal to the sum of factors signs count: $k = \sum_{i=1}^{n} k_i$, where k_i – signs count of factor i.

The result is a system of k linear equations with k unknowns. Values σ^2_{h-ij} are means of variance analysis [5]. σ^2_{rem-11}, σ^2_{rem-12}, …, σ^2_{rem-nk} calculated from the expression (2):

$$\sigma^2_{rem-ij} = \lambda_{ij} - \sum_{x=1}^{n} \sigma^2_{x-ij} \tag{8}$$

Solving the system of equations (7), we can find expected value of karst collapse intensity indicator λ with a combination of various factors using the property of variance (1)

$$\lambda = \sum_{i,j}^{n} \sigma^2_{ij} . \tag{9}$$

3. Determination of karst collapse intensity indicator in area of nuclear power plant construction

Construction project of Nizhny Novgorod nuclear power plant (NPP) is being developed in area, called "Monakovo". Area is located in the basin of the lower reaches of the Oka River. Right tributary of the Oka River is Bolshaya Kutra. River drains the water alluvial, urzhum, lowerkazan and faithfulcoal deposits. Lowland karst formation is due to the large amount of carbonate and sulfate rocks, which lie near the surface and the presence of low-mineralized water in the alluvial layer. Technogenic increase in groundwater level may lead to a more active karst formation and increase the risk of catastrophic situations at NPP.

Assessment of karst hazard of area NPP construction is difficult. Earlier in the territory not carried out comprehensive monitoring of karst processes. Detailed monitoring carried out in the villages Monakovo and Chud. Some data on karst processes are available in the area of agricultural fields and roads. Information about the karst processes in forest and non-agricultural areas almost missing.

Comprehensive observation of the karst processes development in NPP construction has made only a few years ago. This is not sufficient for accurate determination of karst collapse intensity indicator. Thus, we used method for determination of karst collapse intensity indicator using incomplete data.

Research results showed that in the Nizhny Novgorod NPP construction area there is karst hazard. Part of the territory of construct NPP has manifestation of karst activity. During the construction and exploitation nuclear power plants may be a risk of new karst collapses formation. Therefore it is needed make an additional comprehensive study of the territory Nizhny Novgorod NPP construction area.

4. Summary

Described method can be used to determine the intensity of karst collapse indicator using incomplete data. Using the method on the example Nizhny Novgorod NPP construction area allowed evaluating karst hazard.

5. Acknowledgment

The research is supported by grants from Russian Foundation for Basic Research № 13-07-97510.

References

[1] Makeev Z A 1948 *Moscow Conference on Karst* **4**
[2] Sharapov R V 2013 *Fundamental research* **1-2** 444-447
[3] Sharapov R V 2013 *Engineering industry and life safety* **1** 28-34
[4] Tolmachev V V 1986 *Engineering and building development of karst territories* (Moscow, Stroyizdat)
[5] Scheffe H 1959 *The analysis of variance* (Wiley, New York)
[6] Roshan A D, Shylamoni P, Acharya S 2007 *Monograph on siting of nuclear power plants*
[7] Demek J, Kalvoda J 1992 *GeoJournal* **28** 395-402
[8] Fischer J A,. McWhorter J G 2003 *Geotechnical special publication* **122** 485-491
[9] Assaad F A, Jordan H 1994 *Environmental Geology* **23** 228-237

Climate Change, Global Warming and Global Inequity in Developed and Developing Countries (*Analytical Perspective, Issue, Problem and Solution*)

A S Wijaya
School of Earth and Environmental Sciences
James Cook University
Townsville, Queensland, Australia

Email:arifwj5@gmail.com/arif.wijaya@my.jcu.edu.au

Abstract : Climate change becomes one of the most significant challenges faced by most environmentalist all over the world. Every country either developed or developing one has the same need in climate change impact mitigation and adaptation. However, developed countries are believed to have better ability rather than developing countries in particular to climate change adaptation impact. It is described by several indications pointed out by several practitioners. The methods compare findings in both developing and developed countries. It is analyzing two salient data justified by rational arguments and emphasizing with some justifications then finally summarizing with solutions and recommendations.

Keywords: *Developed, developing, Climate change, and Adaptation.*

1. Introduction

In the beginning of the 21st century, Climate change begins to be a new challenge a priority concern of various stake holders. It is taking place due to human activities, and will have significant social, economic and environmental consequences. These consequences will be unevenly distributed across geographical, social and generational boundaries. There are unequal conditions related its mitigation and adaptation impact between developed and developing countries. Development countries or developed countries are countries that already had sufficient economic income and infrastructure facilities and likely to have better mitigation and adaptation strategy dealing with climate change. In contrast, the developing countries just began their development including adapting with climate change Developed nations have adequate preparation with climate change impact owing to their finances and technology, whereas developing countries have limited capability while they become more vulnerable from climate change. This essay will explore how climate change brings global inequity particularly for developing countries then analyze whether developed countries are having better position than developing nations and expel some solutions for climate change impact.

2. Discussion, Methods, and Analysis

Climate change and global warming have more impact on third world nations or developing countries rather than in developed nations. (Shukla 1999) claimed that the distribution of world funding, military stock, knowledge and natural resources is controlled by the developed nations. The power of developed countries in bargaining the world economy had a great influence on the developing countries. It can be argued that developing countries have limits and insufficient financial capacity compare to developed nations. This situation has direct effect on developing countries dealing with climate change impact. To illustrate from the agricultural yields of developing countries, the lack of funding resources, human resources, and technology have been contributing to their sustainable agricultural activities. Nevertheless, Most of Common wealth countries, European countries and The United States have better in those capacities to protect their agricultural production from climate change impact. This indicates the imbalance capacity between developed and developing nations which is not only from one sector like agriculture but also from others such as natural resource management adapting with climate change. However, developed countries can assist developing nations through international cooperation. For example, USAID and AUSAID programs for climate change project in some developing countries.

Developing nations seems to be more vulnerable to climate change due to their economy relying more heavily on climate-reliant activities in particular farming (Fankhauser et al. 2000). In a specific finding, Guldberg reported that storms and coral bleaching as the consequences of climate change activities could potentially destroy fishing grounds and farming systems particularly in the world Coral Triangle (Guldberg et al. 2009). The coral triangle is located at several developing countries in south East Asia including Indonesia. This area supplies more than 50 percent of fish nutrients for world consumption. It can be calculated if this area is continually threated by climate change, then the world community would no longer be able to consume fish. A solution that the developing countries particularly in this area may apply to minimize the effect of coral destruction due to climate change is by adopting the coral conservation in the Great Coral Barrier in Queensland Australia. The regional governments of North Queensland in cooperation with the community who are living in the GBR are have made strong regulation toward the national park of the great barrier. This can be applied in some coral islands which located in the World Coral Triangle. As a result, the fish ecosystem in particular coral fishes in GBR is properly being protected by both the government and the community. Furthermore, ecotourism concept is also applied by the North Queensland Government to conserve the ecosystem values of the Great Barrier Reefs considering its World Heritage Area status.

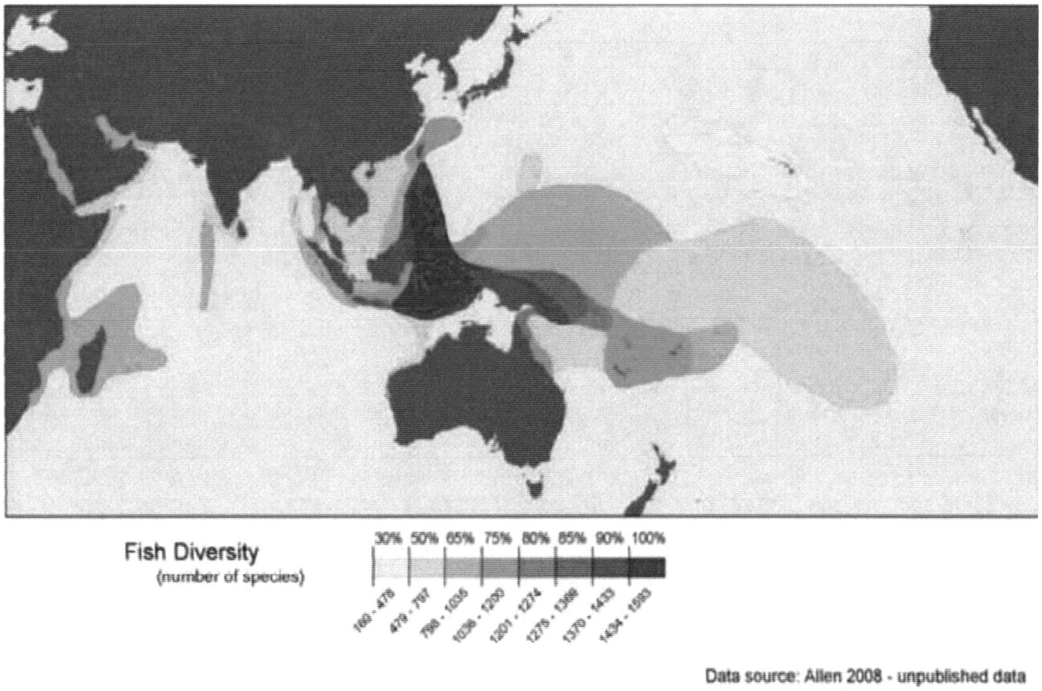

Figure 1. Coral reef fish diversity in the Indo-Pacific Region (Allen 2007, unpublished data). The highest diversity areas (90-100% of species) include most of the Coral Triangle

Other data stated that people in the developing nations have limited access to modern energy (PODOBNIK 2006). Contrarily, the highest rates of energy consumption are founded in developed countries and this is a major cause of global greenhouse gases. Developed nations become rich nations due exploring and utilizing more energy in their industrial activities that is also higher than developing nations. This can be claimed as hegemony of developed nations toward developing countries. Additionally, This condition is discriminating to developing countries especially at international carbon trading system because the developing countries becomes the emission filter for the green houses gases which is excessively produced by the developed countries.

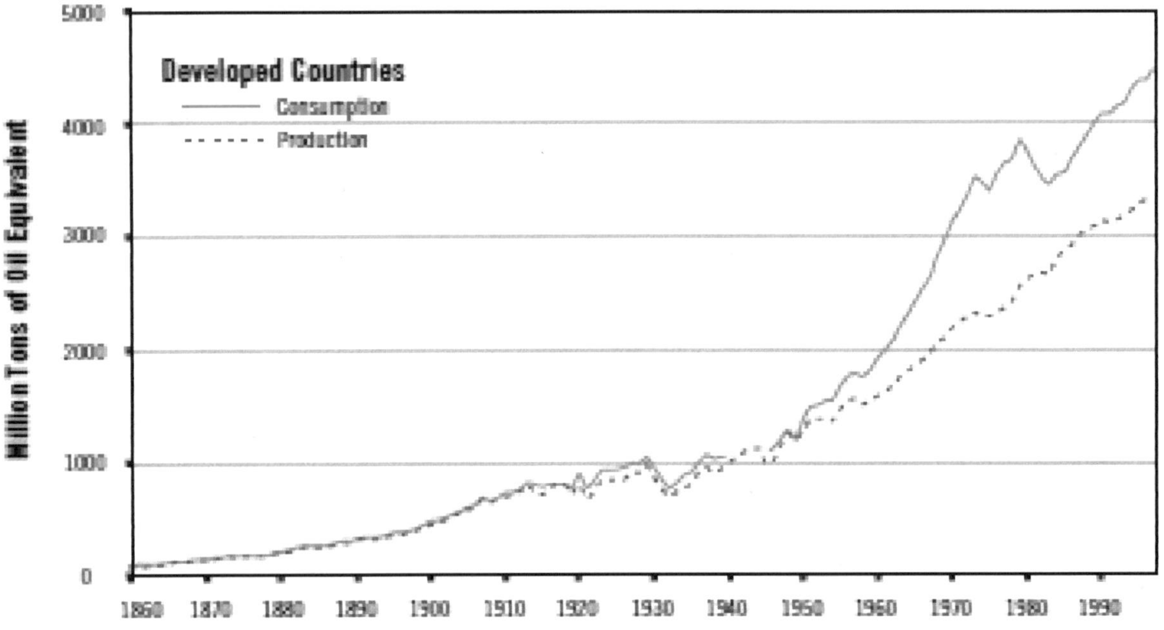

Figure 2. Commercial Energy Production and Consumption, 1860–1998

Despite the fact that developed countries seem to occupy developing countries through some indicators aforementioned, there are a number of opposing arguments. Firstly, the Virgin Group chairman in 21th, September, 2006 said that the company would promise to grant about 1.6 billion that is similar to the income of the Virgin group for 10 years to support the developing countries in environmental program (Prudham 2009). However, there might be a conspiracy in promoting a green movement like in that initiated by the Virgin Group. There are many companies like Virgin group which have already pledged to subsidize their income for environmental conservation including for climate change adaptation and mitigation. In fact, many of those companies have been failed applying their promises in environmental conservation. There should be an independent law institution like the commission for eradication corruption in Indonesia (KPK) that can control, oblige, or even prosecute such companies. The second is several developed nations initiate and support the need to provide global financial assistance in climate change adaptation for developing nations. Yet, most of developing countries are strongly vulnerable affected by the climate change impact due to the countries' population, dynamical profit, natural resources, poor condition, and their land declination

(Smit and Pilifosova 2003).
Actually, the Kyoto Protocol in 1996 has already recommended developed countries to have commitments in alleviating the impact of climate change especially in lessening carbon emission (Harte et al. 2000). Several international agreements have been signed by most of countries, Developed countries such as United States and Australia likely to stay in different position. Albeit climate change and the impact could not be prevented by world communities, either developed countries or developing countries could work together. Developed countries should give financial subsidies in dealing with the climate change impact especially on community's livelihood, for instance in agriculture or aquaculture sectors. In addition, developed countries can empower the capacity of human resources in developing countries by training and courses in climate change mitigation and adaptation. Furthermore, professional climatologist from developed countries could share their knowledge and experiences with scientists in developing countries through cooperation in educational development. Developed countries could provide supporting technology of climate change sciences for developing countries. This is very important to initiate and even strengthen research toward climate change science in developing countries. The research reports would be the salient recommendation for whole stakeholders including the community in developing countries to be able to adapt into the climate change. Whereas the developing countries, for example those which have global forest, would be beneficial for the developed countries that contribute more in global warming.

Although climate change and the impact could not be prevented by world communities, either developed countries or developing countries could work together. Developed countries should give financial subsidies in dealing with the climate change impact especially on community's livelihood, for instance in agriculture or aquaculture sectors. Developed countries can empower the capacity of human resources in developing countries by training and courses in climate change mitigation and adaptation. In addition, professional climatologist from developed countries could share their knowledge and experiences with scientists in developing countries through cooperation in educational development. Developed countries could provide supporting technology of climate change sciences for developing countries. This is very important to initiate and even strengthen research toward climate change science in developing countries. The research reports would be the salient recommendation for whole stakeholders including the community in developing countries to be able to adapt into the climate change. Whereas the developing countries, for example those which have global forest, would be beneficial for the developed countries that contribute more in global warming.

3. Conclusion and Solution

Summing up, it is clear that climate change has more impact on developing countries compared to the developed countries. The impact can be seen from the environmental damages due to climate change that is more significant in developing countries than that in developed and development countries. The developing countries have insufficient funding to apply programs and actions dealing with climate change impact in its regions. Developing countries tend to be fragile and vulnerable in term of climate change impacts and related natural disasters such as drought and storms due to their geographical conditions compare to developed countries. The developed countries consume more global energy and contribute more to global emission than developing countries that cause global inequity in energy consumption in the world and the developing countries contribute in reducing the global carbon emission.

Finally, the impact of climate change and global warming in developing countries absolutely outweighs that of developed countries and this situation has led to global inequity regarding the urgency to give the same attention in climate change action across the world as emphasized in the Kyoto protocol, Copenhagen, and so on. Therefore, developed countries must support developing countries in all solutions related to climate change impact. For instance, by granting funds, providing technology, and assisting through educational development and research in collaboration with scientists between both countries. In some extents, the developed countries can help the community in developing countries dealing with climate change through environmental education programs not only by granting financial support to build schools but also approaching the developing countries to initiate and make education curriculum for environmental education like climate change awareness. This curricula could be taught from primary schools to senior high schools. The developed countries can award scholarships to the young generation of developing countries to study on climate change science in developed countries.

4. References :

[1]. Baer, P., et al. (2000). "Equity and greenhouse gas responsibility." Science **289**(5488): 2287.

[2]. Hoegh-Guldberg, O., et al. (2009). "A comprehensive study involving over 20 experts and based on 300 peer-reviewed scientific articles."

[3]. PODOBNIK, B. (2006). "Global Energy Inequalities Exploring the Long-Term Implications." Global Social Change: Historical and Comparative Perspectives: 135.

[4]. Prudham, S. (2009). "Pimping climate change: Richard Branson, global warming, and the performance of green capitalism." Environment and planning. A **41**(7): 1594.

[5]. Shukla, P. R. (1999). "Justice, equity and efficiency in climate change: a developing country perspective." Fair weather: 145-159.

[6]. Smit, B. and O. Pilifosova (2003). "Adaptation to climate change in the context of sustainable development and equity." Sustainable Development **8**(9): 9.

[7]. Tol, R. S., et al. (2000). "How much damage will climate change do? Recent estimates." WORLD ECONOMICS-HENLEY ON THAMES- **1**(4): 179-206.

Willingness to pay for highlands' agro-tourism recreational facility: A case of Boh Tea plantation, Cameron Highlands, Malaysia

Syamsul Herman, M. A.[1,2*], Nur A'in C. M. [1], Ahmad, S[3] and Ramachandran, S.[1,2]

[1] Faculty of Forestry, Universiti Putra Malaysia, 43400 UPM Serdang, Selangor Darul Ehsan, Malaysia
[2] Institute of Tropical Forestry and Forest Products, Putra Infoport, Universiti Putra Malaysia 43400 UPM Serdang Selangor Darul Ehsan, Malaysia
[3] Institute of Agricultural and Food Policy Studies, Putra Infoport, Universiti Putra Malaysia 43400 UPM Serdang, Selangor, Malaysia
* Corresponding author: syamsulherman@gmail.com, syamsulhma@upm.edu.my

Abstract. The increase in tourist demand for highland experience is inevitable. Cameron Highlands, established as a Tea Plantation Estate during the British Colonial era in 1929, has evolved into a major highland tourism destination providing a cool climatic experience coupled with scenic beauty in the midst of Tudor concept architecture which enhances the destinations historical value. Realising such tourism potential, the Boh Plantation management has provided a visitor centre as recreational facility for tourist utilisation. However, the absence in imposing an entrance fee has left a vacuum in determining the recreational economic value of this facility as the benefit of this agro-tourism product to tourists remains unknown. It would be important for the management to identify the benefit since the development and maintenance of the facility is costly. Hence, the purpose of this paper is to estimate the benefit of such establishment in highlands area by assessing visitor's Willingness to pay (WTP). The study examines, explores and debates the issues in a critical yet supportive environment especially highlands. The study obtained 179 usable questionnaires from visitors during weekends, weekdays and public holidays. The result showed that 59% of the visitors were willing to pay for the agro-tourism product. The WTP was estimated at RM 7.21 (€1.81). Three factors were found to be influencing WTP which were monthly income, years of education and perception on scenery. Although the study was conducted post development, the finding indicated the WTP for current management practise. Should the management change its style, it would also affect WTP and also the total economic value. Since WTP is established concept, the finding of the study reflects on the opportunities, barriers and challenges inherent in embracing post-disciplinary approaches to research and suggest ways to further enhance the approach.

1. Introduction

Recreation activities can be defined as active participation occurs at the source of supply- the site- although elements of the experience- particularly anticipation and reflection- occur back home

(Seabrooke and Miles, 1993). Many recreational activities are related to its environment. Highlands has been attracting tourist to visit for recreational purposes. In Malaysia, highland recreation areas are identified as main tourism destination of the country. One of the famous locations is Cameron Highlands (CH), which is very synonym to outdoor recreational activity, ecotourism and agriculture.
CH is the smallest district in the state of Pahang, Malaysia which is situated at the North-Western of the state (Figure 1). The total of land area is 71, 218 ha. What is interesting about CH is that, it is not only the focus of attention of the visitor either from local people or foreigner; instead it has been recognized as the biggest highland resort in Malaysia. Among the attraction of this destination is that the location is located at the Titiwangsa Mountain Range with the elevation of 1500 meters from the sea level. Due to its altitude, the temperature in CH varies between 18° C to 25 °C during the day. CH is also known for montane climate vegetation and fauna, together with many beautiful landscapes. Many infrastructures in CH have adopted the English-style villages with various flowers garden, due to its historical development.

Table 1. Visitor Arrivals to BOH Tea Plantation, Cameron Highlands

MONTH	YEAR 2011	YEAR 2010
January	10450	11159
February	10450	10431
March	10450	16729
April	10450	14882
May	10450	15196
June	10450	23731
July	10450	25147
August	10450	11677
September	10450	17236
October	10450	20839
November	10450	19363
December	10450	26937
TOTAL	**125400**	**213327**

2. BOH Tea Plantation, Cameron Highlands as a Tourism Area

One of the tourism destinations in CH is the BOH Tea Plantation (BTP) located in Sg. Palas. The history of BTP started by an Englishman, A.J. Russell in 1929, with its original purpose was mainly for tea production. The plantation has been a main tea producer in the country since then. BTP covers an area of 8000 acres and solely planted with tea and has been recognized as the biggest tea plantation in Southeast Asia. Nowadays, BTP has become one of a main tourist attraction in CH.
Realising the growth in visitor demand, BTP has erected a visitor centre in Sg. Palas, housing an indoor amphitheatre, a teashop, washroom, a café with vista, parking and interpretive exhibits. In this centre, a free guided factory tours on tea production and process factory is made available for visitors. Statistics in year 2010 and 2011 (Table 1) shows an increasing visitor arrivals to BTP. This shows the popularity of BTP as a recreational area at CH. This place is always crowded on weekends and school holidays. This facility requires high cost in development and the maintenance of the facilities due growing number of tourist arrivals. Since there is no entrance fee, it is a good

practice for BTP management to understand of how much the facility has benefited their visitors. Hence, the purpose of the study is to estimate the willingness to pay (WTP) of local visit to BTP, CH.

The significant of this study to the BTP's management is that the study can reveal the value associated with current management scheme. This information is useful in planning especially in relation to budget allocation where the estimated value can justify the allocation for development and maintenance of the area. In addition, this research also provide visitor's typology, which means the study gives more information about travel characteristic on visitors to the managing agency and this may help the agencies in understanding their visitors.

3. Contingent Valuation Method and Willingness to Pay

Contingent Valuation Method (CVM) is one of the method that used to estimate the economy value which have in a certain area that involve service and also the existing source of nature. There were a lot of opinion about this method that have been stated in the previous research with one of them is from Ajzen and Driver (1992) which stated that CVM is a tool used to place an amount of money for goods and service that usually do not have place in market (Zaiton et al., 2012). There are also people stated that CVM is a technique used to evaluate all types of ecosystem and service that related to the environment. This method was used to assess neither direct usage nor indirect usage.

In CVM research, users will be directly asked on how much money they are willing to pay for services that related with the environment. In CVM, the main concept use in WTP is the maximum value that people are willing to pay for towards goods or for services. Particularly, the WTP is an amount of money that people are willing and able to invest to ensure they can enjoy the source and the recreation facilities provided to them (McConnell, 1985). Other than that, Dupont (2002) said that in order to obtain WTP measures for environmental quality improvements leading to better recreational opportunities, the payment of a specific amount of money is arrange to fund the activities that will lead to the outcome ascribed to survey respondents.

CVM has been tremendously used in assessing users' recreational benefit studies. Yuan (2009) conducted a study on the visitor's WTP for facilities and services in Kuala Lumpur City Center (KLCC) Park, Malaysia. He estimated that the minimum amount that respondent agree to pay was RM1.00; while the maximum amount was RM30.00, and the mean for this maximum amount was RM5.40. Hayati (2009) estimated the WTP for urban forest in Johor Bahru, Malaysia was as low as RM1.67. This study adopted CVM as the framework in order to obtain visitors' WTP.

4. Methodology

4.1 Sampling and Fieldwork procedure

Data for this study were collected using a structured questionnaire. The questionnaires were distributed to the visitors in BTP. The data collection period covers weekdays, weekends, public holidays and school holidays. A total of 180 useable questionnaires were obtained during 3 months data collection in 2012. The questionnaire consisted of 3 sections; section 1 on visit and travel characteristics; section 2 on WTP; and section 3 on visitor profile. To ensure a full returned questionnaire, visitors were interview face-to-face by trained enumerators. Data collected were analyzed using SPSS version 20.0.

4.2 WTP Model

In the study, respondents were asked if they were willing to pay if the management charge entrance fee to the visitor centre of BTP. Respondents were explained that the fee was for maintenance of the recreational facility and also for conservation purposes. The study adopted modified bidding game technique. In this technique, the respondents were approached with different starting WTP prices. If a respondent agrees with proposed price, they were asked to state the highest WTP above the proposed price. On the contrary, if the respondents do not agree with the proposed price, they were asked to state the highest WTP price below the proposed price. To determine the WTP price and factor influencing WTP, a multiple regression was employed. The functional model is shown as in Equation 1 below;

$$WTP = f(VISIT + ONSTIME + SIGHT + UNIQ + BEAU + RECM + ATTR + EXP + AGE + DIST + YSREDU + SI + INC + C) \quad \ldots (Eq.\ 1)$$

Where:

WTP	- Maximum amount of WTP (Dependent variable)
VISIT	- Frequency of visits for the past 12 months (Scale data)
ONSTIME	- Visitors' length of stay in term of hours
SIGHT	- Sightseeing activities as a motive of visit (Dummy, 1=yes, 0=no)
UNIQ	- Uniqueness of Topography as a motive of visit (Dummy, 1=yes, 0=no)
BEAU	- Perception of beautiful scenery as a motive of visit (Dummy, 1=yes, 0=no)
RECM	- Recommendation from friends as a motive of visit (Dummy, 1=yes, 0=no)
ATTR	- Attractive promotion as a motive of visit (Dummy, 1=yes, 0=no)
EXP	- Total of expenditure on site (Scale data)
AGE	- Age of respondent (Scale data)
DIST	- Distance from residential area to BTP (Scale data postcode)
YSREDU	- The respondents' number of year educational level (Scale data)
SI	- Respondents' satisfaction of recreational activities, facilities and services provided (Likert scale, 1= strongly disagree, 2= disagree, 3= agree, 4= strongly agree)
INC	- Monthly income (RM/month)
C	- Constant

5. Results and Discussion

5.1 Visitor Profile
Fifty seven percent of the respondents are female and 51% are single, suggesting almost equal visitors profile to BTP. Majority of the respondents (79.9%) had tertiary education level. Mean income is found to be RM3,235.15 (€743.46). The mean for age is 30 years old (Table 2). These suggest that respondents to BTP are mainly educated, young adults with middle income.

5.2 Visit Characteristics
The average frequency of visiting to BTP is 1.79 times per year. 55.9% respondents stated that it is their first time visiting BTP. Average time spent at the site is 2.36 hours. 45% respondents come with their family. Majority of the respondents (76%) indicated that the purpose of visit is to enjoy the scenery and environment at BTP. The result suggests that the visitors to BTP are day-trippers, new and stay only for a short while. They visited BTP to enjoy the natural environment at the site.

5.3 Willingness To Pay for Recreational Experience and Conservation
The result shows that 59% respondents said that they are willing to pay if the management charges the entrance fee is for the maintenances and conservation of the area.

Table 2. Socio-Economic Statistics

	Minimum	Maximum	Mean	Std. Deviation
Age	18	65	30.60	8.857
Year Education	11	17	15.83	2.386
Monthly Income	RM500 (€114.90)	RM30,000 (€6,894.23)	RM3235.14 (€743.46)	2965.48

5.4 Regression Analysis Result

The determination coefficient, R^2 measures the proportion of the variation in the dependent variable explained by the independent variables. A higher R^2 would imply that the calculated Y equation line fits closer to the data points (Ahmad, 1994). However, the R^2 for the study was found to be low (0.163), which showed that only 16.3% of the respondents' WTP is explained by the independent variables. One possible reason is due to the selection of independent variables used in the model. There may be other possible variables which may have influences on visit decision which are not included in the model.

T-test at 95% confidence interval (t-tabulated value 1.96) was employed to test for statistical significance of regression coefficients. Any t- computed value which is greater than the t-tabulated value implies that the variable plays an important role in determining the dependent variable. From the 14 factors that have been listed, only three of the factors were significant at 0.05 confidence level (Table 3). These factors are beautiful scenery, years of formal education and monthly income. The coefficient indicated that the relationship between independent variable and dependent variable either directly proportional or inversely proportional. The result shows that beautiful scenery and monthly income are directly proportional with maximum amount of WTP; while the year of formal educational is inversely proportional with maximum amount of WTP.

To interpret the results, the variable perception on beautiful scenery has influenced visitor's WTP positively. The more they enjoy the scenery, the more they are willing to pay. This is consistent with the main reason of visiting BTP; to enjoy the natural environment of BTP. BTP is well known for its beautiful scenery of tea plantation at low temperature climate. Therefore, it is a popular destination among visitors especially during the school holidays and weekends. In other words, they are willing to spend more money if they feel they can enjoy the scenery more; in a longer period or a better quality.

Monthly income is found to be influencing the WTP and it is directly proportional with WTP by the visitors to that place (b=0.001). This is because as income increase; the WTP also increase due to higher purchasing power. This statement is supported by Yuan (2009), which stated that the higher the income, they are more WTP and the reason is because they have more discretionary income to enjoy recreational activities that they participated.

The third variable is the years of formal educational. It is found to be inversely proportional to WTP (b= -0.513). Demographic analysis shows that most of them are educated, hence making the visitors sensitive and aware towards of environmental issues such as land clearing for plantation, landslides and etc. CH is also well known for large areas of vegetable farming. Many of these farms are located at slope areas. There are environmental degradation issues such as soil erosion and landslides in CH in relation to human activities such as farming. Thus, visitors with higher education level have more awareness. They are more sensitive towards environmental intervention issues and quality degradation. It is assumed that these are the factors that contribute to the negative relationship between education and WTP.

Table 3. Factors Affecting Visitor's WTP for Recreational Experience at Sungai Palas BOH Tea Plantation, Cameron Highlands

Variable	B	Beta	p-value
(Constant)	4.797		0.467
Beautiful Scenery	**2.929**	**0.167**	**0.025***
Year of Education Level	**-0.513**	**-0.178**	**0.027***
Monthly Income	**0.001**	**0.229**	**0.005***
Frequency Visiting	-0.225	-0.076	0.379
On site Time	0.402	0.060	0.421
Sightseeing Activity	0.018	0.001	0.988
Uniqueness of Topography	-2.006	-0.137	0.074
Recommendation From Friends	-1.180	-0.056	0.453
Attractive Promotion	1.023	0.033	0.670
Total Expenditure	0.001	0.078	0.324
Distance	-0.008	-0.096	0.272
Age	-0.009	-0.012	0.877
Satisfaction Index	1.187	0.070	0.362
Crowding	0.891	0.100	0.188

a. Dependent Variable: Maximum amount (WTP)

5.5 Estimated Mean WTP

Based on the three identified variables influencing WTP, a mean WTP is then estimated using Equation 2 below:

$$\mathbf{WTP} = f(\mathbf{BEAU + YSREDU + INC + C}) \quad ...(Eq.2)$$

Taking the mean value of all independent variable, the mean WTP value is estimated at RM7.21 (€1.65) for the year 2012. This is the value estimated by taking the factors that affecting the WTP.

6. Conclusion

The BOH tea plantation has become one main attraction not only in Cameron Highlands but in the country as a whole. The attraction in the tea plantation greatly depends on the naturalness of the area. It is the main reason of visit arrival, the conservation of its natural environment is the utmost importance. Hence, the study is conducted to assess how much the visitors are willing to contribute to recreation and nature conservation. The study has found that visitors are willing to pay for recreational facilities and conservation at BTP. Nevertheless, the WTP value is relatively low.

It is recommended that subsequent study to be conducted to identify the low WTP of domestic tourists. A comparative study on foreign tourist is also suggested although number of foreign tourist is slightly low than domestic tourists. From the visitors' profile it is understood that most of them are educated with middle income and accompanied either by family or relatives. It is, therefore suggested more educational programmes and facilities to be provided at the location. This might enhance visitors experience at the location and might contribute to their satisfaction, thus increasing the WTP. The study has provided to an indicator of visitors' satisfaction to the management of BTP.

7. Acknowledgment
The authors would like to express their gratitude to the management of BOH Tea Plantation for their permission and assistance during the study.

8. References
Ahmad S 1994 *Demand for and value of outdoor recreation in Langkawi by domestic visitors* Ph.D. dessertation Universiti Putra Malaysia

Ajzen, I and Driver B L 1992 Application of the theory of planned behaviour to leisure choice *Journal of Leisure Research* **24** 207- 224

Dupont D P 2002 *Gender and willingness-to-pay for recreational benefits from water quality improvements* Canada: Brock University

Hayati N S 2009 *Kesanggupan membayar oleh pengunjung terhadap Hutan Bandar di Daerah Johor Bharu, Johor* Degree thesis Universiti Teknologi Malaysia

McConnell K E 1985 Valuation congested recreation sites *Journal of Environmental Economics and Management* **7** 389-394

Seabrooke W and Miles C W N 1993 *Recreation land management* USA: University of Reading

Yuan L P 2009 *Visitors' willingness to pay for facilities and services in Kuala Lumpur City Center (KLCC) Park* Degree thesis Universiti Putra Malaysia

Zaiton S Syamsul Herman MA Alias R and Mohd Rusli Y 2012 Willingness to pay for conservation fee at Penang National Park *Malaysian Forester* **75** 1 43-52

2013 International Conferences on Geological, Geographical, Aerospace and Earth Sciences IOP Publishing
IOP Conf. Series: Earth and Environmental Science **19** (2014) 012010 doi:10.1088/1755-1315/19/1/012010

Free vibration of symmetric angle-ply laminated circular cylindrical shells

K K Viswanathan[1], Zainal Abdul Aziz, H Z Amirah, Saira Javed

UTM-Center for Industrial and Applied Mathematics (UTM-CIAM)
Department of Mathematical Sciences, Faculty of Science,
Universiti Teknologi Malaysia, 81310 Skudai, Johor Bahru, Johor, Malaysia
e-mail: visu20@yahoo.com, viswanathan@utm.my

Abstract. Free vibration of symmetric angle-ply laminated circular cylindrical shells is studied using Spline approximation. The equations of motions in longitudinal, circumferential and transverse displacement components, are derived using Love's first approximation theory. The coupled differential equations are solved using Spline approximation to obtain the generalized eigenvalue problem. Parametric studies are performed to analyse the frequency response of the shell with reference to the material properties, number of layers, ply orientation, length and circumferential node number and different boundary conditions.

1. Introduction

Shell structures are widely used as structural components in engineering, aerospace, naval and chemical industry. They are often used as load-bearing structures for aircrafts, rockets, submarines and missile bodies. The cylindrical shells used in those designs are stiffened to achieve better strength, stiffness, temperature resistant and light weight characteristics. Recently, '[1]' studied laminated shells. Free vibration of layered cylindrical shells of variable thickness was examined and the frequencies were analyzed for specially orthotropic materials using collocation with splines '[2]'. The paper on angle-ply laminated cylindrical shells based on the Love-type version of a unified shear deformable shell theory was studied by '[3]'. A refined high-order global-local theory was used to analyse the laminated composite beams using the finite element method including shear deformation '[4]'. Recently, Topal and Uzman '[5]' analyzed thermal buckling load optimization of angle-ply laminated cylindrical shells. Recently, '[6]' investigated the free vibration of symmetric angle-ply cylindrical shells of variable thickness.

Most of the researchers analyzed the problems of free vibration for angle-ply laminated shells without using spline approximation. Thus, there may not be available sufficient work on angle-ply laminated cylindrical shells with orthotropic materials using spline approximation technique in the past literature. This paper studies the free vibrations of angle-ply layered circular cylindrical shells using spline approximation where the problem is formulated by extending Love's first approximation theory on homogenous shell. The shell is made up of uniform layers of isotropic or specially orthotropic materials.

[1] Corresponding author: E-mail: visu20@yahoo.com, viswanathan@utm.my

Studies are carried out for cylindrical shells with clamped-clamped and simply supported boundary conditions along the axial direction and the layers of the material are considered to be thin .The layers are perfectly bounded together to move without interface slip. The effects of the angle-ply, different materials and geometric parameters on the frequencies of angle-ply laminated cylindrical shells under different boundary conditions are analysed. In the analysis, spline approximation technique will be adopted to approximate the displacement functions to analyse the frequencies.

2. Formulation of the Problem

Consider a thin laminated circular cylindrical shell of having the radius r, length ℓ and thickness h. Each layer in the laminated composite is assumed to be homogeneous, linearly elastic, and specially orthotropic. The layers are perfectly bonded at the interfaces. The coordinate system (x, θ, z) is defined at the mid surface of the shell and u, v and w are the displacements in the directions of x, θ and z respectively.

The equations of motion for thin circular cylindrical shells are given by the following (Viswanathan *et al.*, 2010):

$$\frac{\partial N_x}{\partial x} + \frac{1}{r}\frac{\partial N_{x\theta}}{\partial \theta} = \rho h \frac{\partial^2 u}{\partial t^2} \tag{1}$$

$$\frac{\partial N_{x\theta}}{\partial x} + \frac{1}{r}\frac{\partial N_{\theta}}{\partial \theta} + \frac{1}{r}\frac{\partial M_{x\theta}}{\partial x} + \frac{1}{r^2}\frac{\partial M_{\theta}}{\partial \theta} = \rho h \frac{\partial^2 v}{\partial t^2} \tag{2}$$

$$\frac{\partial^2 M_x}{\partial x^2} + \frac{2}{r}\frac{\partial^2 M_{x\theta}}{\partial x \partial \theta} + \frac{1}{r^2}\frac{\partial^2 M_{\theta}}{\partial \theta^2} - \frac{1}{r} N_{\theta\theta} = \rho h \frac{\partial^2 w}{\partial t^2} \tag{3}$$

The stress resultants and stress couples are given by

$$\left(N_x, N_\theta, N_{x\theta}\right) = \int_z \left(\sigma_x, \sigma_\theta, \sigma_{x\theta}\right) dz, \left(M_x, M_\theta, M_{x\theta}\right) = \int_z \left(\sigma_x, \sigma_\theta, \sigma_{x\theta}\right) z\, dz \tag{4}$$

where N_x , N_θ , $N_{x\theta}$ are stress resultants, M_x , M_θ , $M_{x\theta}$ are moment resultants and σ_x, σ_θ, $\sigma_{x\theta}$, are stresses in respective directions.

For a thin laminated cylindrical shell, the stress and strain relation of the kth layer is defined as

$$\left\{\sigma^{(k)}\right\} = \left[Q^{(k)}\right]\left[\varepsilon^{(k)}\right], \tag{5}$$

where $Q^{(k)}$ are elastic coefficients and $\varepsilon^{(k)}$ are normal and shear strains of the k-th layer.

When the materials are oriented at an angle θ with the x-axis, the transformed stress–strain relations are given by

$$\left\{\sigma^{(k)}\right\} = \left[\overline{Q}^{(k)}\right]\left[\varepsilon^{(k)}\right], \tag{6}$$

where $[\overline{Q}^{(k)}] = [T]^{-1}[Q^{(k)}][T]$ is defined in '[7]'.

Substituting (6) into (4), we get the equations of stress resultants and moment resultants as follows.

$$\begin{bmatrix} N \\ M \end{bmatrix} = \begin{bmatrix} A_{ij} & B_{ij} \\ B_{ij} & D_{ij} \end{bmatrix} \begin{bmatrix} \varepsilon \\ \kappa \end{bmatrix} \tag{7}$$

where $\varepsilon_x = \partial u/\partial x$, $\varepsilon_\theta = (1/r)\partial v/\partial\theta + w/r$, $\varepsilon_{x\theta} = \partial v/\partial x + (1/r)\partial u/\partial\theta$, $\kappa_{xx} = -\partial^2 w/\partial x^2$,
$\kappa_{\theta\theta} = (1/r^2)\partial v/\partial\theta - (1/r^2)\partial^2 w/\partial\theta^2$ and $\kappa_{x\theta} = (1/r)\partial v/\partial x - (2/r)\partial^2 w/\partial x\partial\theta$ (8)

Here u, v and w are the displacement functions of the mid plane in longitudinal, circumferential and transverse directions and A_{ij}, B_{ij} and D_{ij} are the laminate stiffnesses defined by

$$A_{ij}=\sum \overline{Q}_{ij}^{(k)}(z_k - z_{k-1}),\ B_{ij}=\sum \overline{Q}_{ij}^{(k)}(z_k^2 - z_{k-1}^2),\ D_{ij}=\sum \overline{Q}_{ij}^{(k)}(z_k^3 - z_{k-1}^3),\ i,j=1,2,6. \quad (9)$$

Here z_k is the distance of the k-th layer from the reference surface.

It is assumed that there are no stretching-shearing, twisting-shearing and symmetric angle-ply lamination, so that $A_{16} = A_{26} = D_{16} = D_{26} = 0$ and all $B_{ij} = 0$. Substituting ' equation (7)' in to the 'equations (1-3)' one can obtain the differential equations in terms of displacement functions u,v and w.

The displacement components u,v and w are assumed in the separable form given by

$$u(x,\theta,t)=U(x)\cos n\theta\, e^{i\omega t},\ v(x,\theta,t)=V(x)\sin n\theta\, e^{i\omega t} \text{ and } w(x,\theta,t)=W(x)\cos n\theta\, e^{i\omega t} \quad (10)$$

where x and θ are the coordinates defined in the longitudinal and circumferential directions respectively, ω is the angular frequency of vibration, t is the time and n is the circumferential node number.

Substituting the 'equation (10)' in the coupled differential equations, one can obtain the equations in terms U, V and W and can be written in the matrix form as

$$\begin{bmatrix} L_{11} & L_{12} & L_{13} \\ L_{21} & L_{22} & L_{23} \\ L_{31} & L_{32} & L_{33} \end{bmatrix} \begin{bmatrix} U \\ V \\ W \end{bmatrix} = [0] \quad (11)$$

where L_{ij} are linear differential operators in X.

The differential equations on the displacement function contain derivatives of third order in U, second order in V and forth order in W. As such they are not amenable to the solution procedure. Hence the equations are combined within themselves and a modified set of equations are derived. The procedure adopted to this end is to differentiate the first of 'equation (11)' with respect to x once and to use it to eliminate U''' in the third equation. The modified set of equations are of order 2 in U, order 2 in V and order 4 in W.

2.1. Non-Dimensionlization

The following non dimensional parameters are introduced:

$\lambda=\ell\omega(R_0/A_{11})^{1/2}$, a frequency parameter;

$\delta_k=h_k/h$, the relative thickness of the k-th layer;

$H = h/r$, the thickness parameter; $L = \ell/r$, a length parameter;

$X = x/\ell$, a distance co-ordinate; $R = r/\ell$, a radius parameter (12)

where ℓ is the length of the cylinder, r is its radius, h_k is the thickness of the k-th layer, h is the total thickness of the shell, R_0 is the inertial coefficient and A_{11} is a standard extensional rigidity coefficient. Clearly the range of X is lies between 0 and 1.

2.2. Method of Solution

The displacement functions $U(X)$, $V(X)$ and $W(X)$ are approximated by the cubic and quantic splines as stated below:

$$U(X) = \sum_{i=0}^{2} a_i X^i + \sum_{j=0}^{N-1} b_j (X - X_j)^3 H(X - X_j), V(X) = \sum_{i=0}^{2} c_i X^i + \sum_{j=0}^{N-1} d_j (X - X_j)^3 H(X - X_j) \text{ and}$$

$$W(X) = \sum_{i=0}^{4} e_i X^i + \sum_{j=0}^{N-1} f_j (X - X_j)^5 H(X - X_j) \tag{13}$$

Here $H(X-X_j)$ is the Heaviside step function and N is the number of intervals into which the range [0,1] of X is divided. The points $X=X_s=s/N$ $(s=0,1,2,...N)$ are chosen as the knots of the splines, as well as the collocation points. By applying the collocation points to the differential equations given in (11) , a set of 3N+3 homogeneous equations in 3N+11 unknown spline coefficients a_i, b_j, c_i, e_k, f_j (i=0,1,2; k=0,1,2,3,4; j=0,1,2, ...N-1), are obtained. To get 8 more equations, we consider the boundary conditions as follows:

(i) (*C*- *C*): both the ends clamped and (ii) (*S* - *S*): both the end are simply supported.

Combining any one of the boundary conditions, resulting a generalized eigenvalue problem as follows:

$$[M]\{q\} = \lambda^2 [P]\{q\} \tag{14}$$

where [*M*] and [*P*] are matrices of order (3*N*+7) (3*N*+7) , $\{q\}$ is a column matrix of order (3*N*+7)×1 and λ is the eigenparameter.

2. Results and Discussions

'Table 1' depicts comparative study made for homogeneous cylindrical shells of circumferential node number n and the natural frequencies obtained for the shell with C–C boundary conditions and compared with the results with the values of the natural frequencies obtained by '[8-12]' for the axial mode values of m=1,2,3 and the circumferential mode values for n=2,4,6 are compared with the corresponding results obtained by the present method. The parametric values used are ℓ =511.2mm, r=216.2mm, h=1.5mm, E=1.83×10^{11} N/m^2, ν=0.3, ρ=7492kg/m^3 '[13]' . The agreement of the current results are very good. The slightly higher differences with the other results quoted must be due to the differences in the formulations, methods of solution and numerical accuracy involved.

Table 1. Comparison of Natural Frequencies for Cylinder Shells with Clamped Edges

Mode		Natural frequencies in Hz					
Axil mode m	Circumferential node n	Present Method	Viswanathan and Navaneethakrishnan '[8]'	Arnold and Warburton [12]	Smith and Haft [11]	Au-Yang [10]	Goncalves and Ramos [9]
1	2	1312	1299	1240	1429	1485	1405
1	4	3920	3872	3970	4142	4147	4004
1	6	9195	9125	9230	9400	9396	9251
2	2	2530	2660	2440	2682	2852	2904
2	4	4123	4043	4160	4335	4350	4214
2	6	9283	9276	9380	9840	9533	9384

'Figure 2' describes the effect of length parameter L with the angular frequencies ω_m ×10^3 *Hz* (m=1,2,3) for three layered shells having ply orientation of 30°/0° /30°, 40°/0° /45° and 60°/0° /60° and with arranging materials in the order of S-glass epoxy (SGE) – High strength graphite epoxy (HSG)- S-glass epoxy (SGE) under *C-C* boundary conditions fixing thickness H=0.015 and circumferential node number $n = 4$. It is observed from the figure, that ω_m decreases with L increasing. The decrease is fast for very short shells ($0.5 < L < 0.75$). Here ω_m decreases rapidly for the value of L between 0.75 and 0.85 after which the decrement is very low for L >0.85. The nature of variation is same for all the modes and the values are higher for higher modes. The per cent changes in ω_m (m=1,2,3) over the range $0.5 < L < 2$ are 276.66%, 317.79%, 449.57% for $30^0/0/30^0$, 254.37%, 266.61%, 373.70% for $45^0/0/45^0$, and 269.74%, 282.63%, 371.51% for $60^0/0/60^0$ respectively.

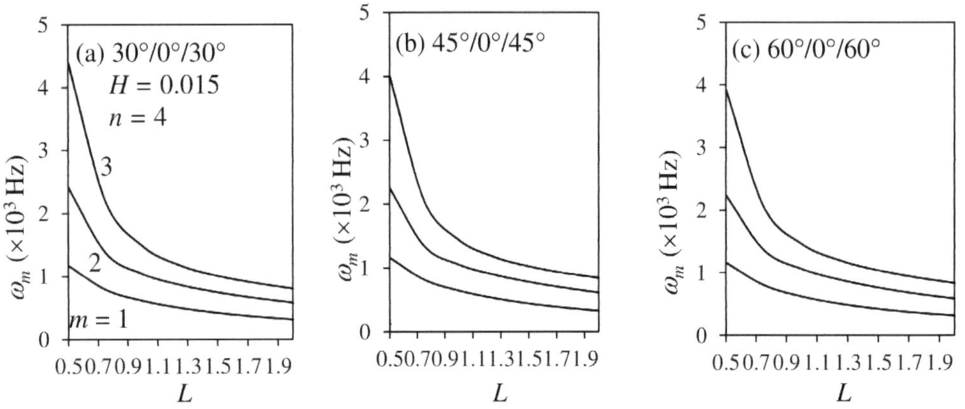

Figure 2. Effect of length of three layered shell on angular frequencies for $n = 4$ under $C-C$ boundary conditions.

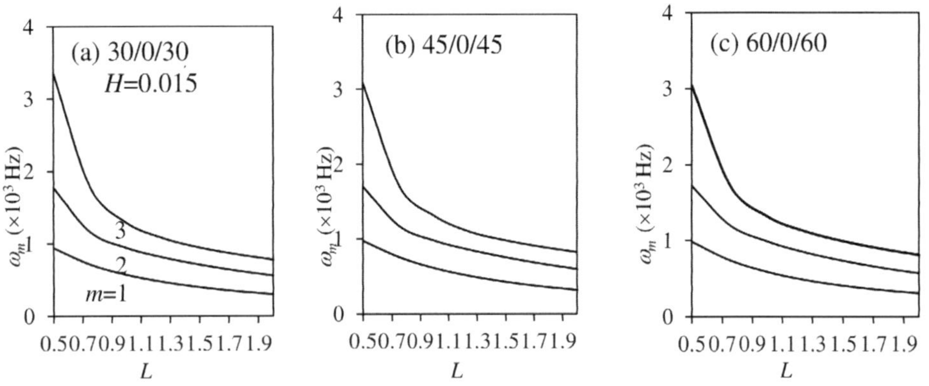

Figure 3. Effect of the length of three layered cylindrical shells with angular frequencies under $S-S$ boundary conditions.

The effect of length parameter L on the angular frequency parameter $\omega_m \times 10^3$ Hz(m=1,2,3) under $S-S$ boundary conditions for three layered shells with ply orientation of 30°/0°/30°, 45°/0°/45° and 60°/0°/60° with the order of the materials in the order of SGE-HSG-SGE are shown in 'figure 3'.

'Figure 4' describes the variation of angular frequencies $\omega_m \times 10^3$ Hz(m=1,2,3) with respect to length parameter L for five layered plates arranging the materials in the order of SGE-HSG-SGE-HSG-SGE with ply angles of 30°/45°/0° /45° /30° under C-C and S-S boundary conditions by fixing H=0.015 and n=4. As

L increases, ω_m is observed to be decreased in general. The decrease is fast for very short shells ($0.5 < L < 0.75$). There is a rapid change in the rate of decrease of ω_m in the $0.75 < L < 0.85$after which the decrease is very low. The value of angular frequency ω_m is higher for higher modes. The per cent changes in ω_m (m=1,2,3) over the range $0.5 < L < 2$ for $30^0/45^0/0/45^0/30^0$ ply orientation under C-C boundary conditions are 253.24%, 276.82%, 349.83% and 174.11%, 230.82%, 254.62% for S-S boundary conditions.

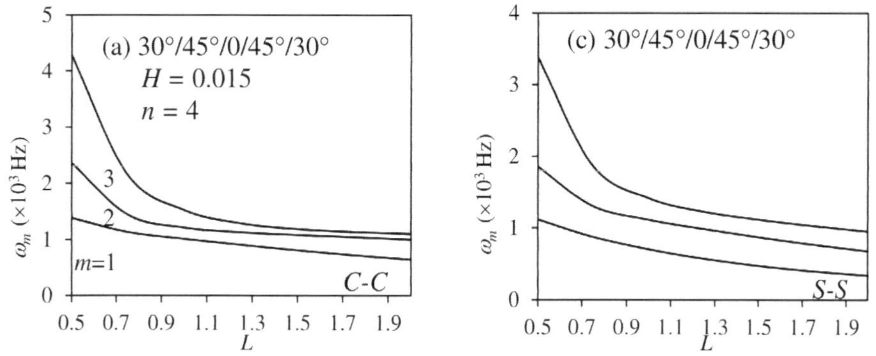

Figure 4. Effect of the length of five layered cylindrical shells with angular frequencies under C-C and S-S boundary conditions.

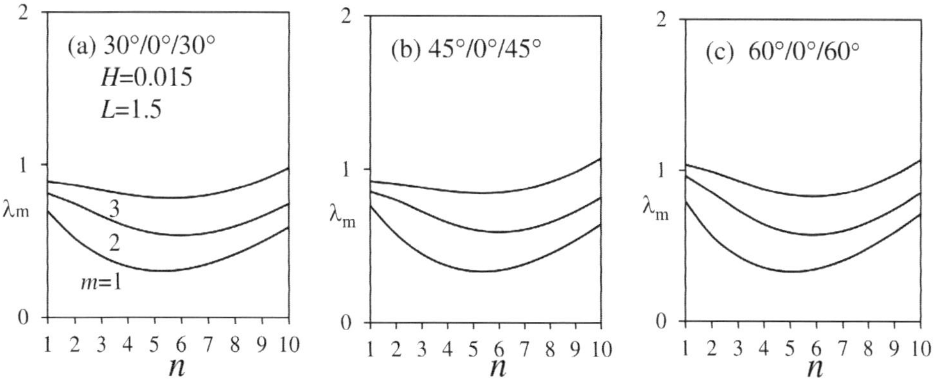

Figure 5. Variation of frequency parameter with respect to the circumferential node number for three layered shells under $C-C$ boundary conditions.

'Figure 5' shows the influence of the circumferential node number n of the frequency parameter λ_m (m=1,2,3)for three layered shells with ply orientation of 30°/0°/30°, 45°/0°/45° and 60°/0°/60° with materials in the order of SGE-HSG-SGE fixing thickness H=0.015 and length L=1.5 under C-C boundary conditions. The per cent changed in λ_l for three cases considered are (a) 9.04%, (b) 3.11% and (c) 5.53%. It is seen at the outset that n affect λ_m significantly. The effect is higher for lower meridional mode values. In general, λ_m decreases with n up to some value of n and then increases which occurs approximately in the range $4< n <6$.

3. Conclusions

Layering of the shell wall influences the natural frequencies of the vibration of the cylindrical shell. The relative thickness, as well as the angle ply orientation between them affects the frequencies. A desired frequency of vibration, within a range of frequencies, may be obtained by a proper choice of the relative thickness of layers and the angle-ply rotation among the chosen materials of the layers. We can choose the desired frequency of vibration from the results by a proper choice of the coefficient of thickness variations and arrangements of ply-angles.

4. Acknowledgement

The authors thankfully acknowledge the financial support from the UTM-Flagship Research Grant Vote No. 01G40, Research management Centre (RMC), Universiti Teknologi Malaysia, Malaysia for completion of this research work.

REFERENCES

[1] Qatu M S 2004 *Vibration of Laminated Shells and Plates* (Netherlands: Elsevier)

[2] Viswanathan K K, Kim K S, Lee J H, Koh H S, and Lee J B 2008 Free Vibration Of Multi- Layered Circular Cylindrical Shell With Cross-Ply Walls, Including Shear Deformation By Using Spline Function Method. *J. Mech. Sci. Technol.* **22** 2062-2075.

[3] Timarci T and Soldatos K P 2000 Vibrations of angle-ply laminated circular cylindrical shells subjected to different sets of edge boundary conditions. *Journal of Engineering Mathematics.* **37** 211-230

[4] Nazargah M L, Shariyat M and Behesti-Aval SB 2011 A refined high-order global-local theory for finite element bending and vibration analyses of laminated composite beams. *Acta Mech.* **217** 219-242

[5] Topal U and Uzman U 2009 Thermal buckling load optimization of angle-ply laminated cylindrical shells. *Materials & Design.* **30(3)** 532-536

[6] Viswanathan K K, Lee J, Zainal A A and Hossain I 2011 Free vibration of symmetric angle- ply laminated cylindrical shells of variable thickness. *Acta Mechanica.* ***221*(3)** 309-319

[7] Viswanathan K K and Lee S K 2007 Free vibration of laminated cross-ply plates including shear deformation by spline method. *Int. J. Mech. Sci.* **49** 352-363

[8] Viswanathan K K and Navaneethakrishnan P V 2003 Free vibration study of layered cylindrical shells by collocation with splines. *Journal of Sound and Vibration* **260(5)** 807-827

[9] Goncalves P B and Ramos N R S S 1997 Numerical Method for Vibration Analysis of Cylindrical Shells. *American Society of Civil Engineers Journal of Engineering Mechanics.* **123** 544–550

[10] Au-Yang M K 1978 Natural Frequency of Cylindrical Shells and Panels in Vacuum and In a Fluid. *Journal of Sound and Vibration.* **57** 341–355

[11] Smith B L and Haft E E 1968 Natural Frequencies of Clamped Cylindrical Shells. *Journal of Aeronautics and Astronautics.* **6** 720–721

[12] Arnold R N and Warburton G B 1949 Flexural Vibrations of the Walls of Thin Cylindrical Shells Having Freely Supported Ends *Proceeding of the Royal Society of London.* **197** 238–256

[13] Chung H 1981 Free Vibration Analysis of Circular Cylindrical Shells *Journal of Sound and Vibration.* **74** 331–350

IOP Conf. Series: Earth and Environmental Science **19** (2014) 012011 doi:10.1088/1755-1315/19/1/012011

Analysis of Land Deformation on Slope Area using PS InSAR. Case Study: Malang Area

Dodi Sudiana[1,4], Rokhmatuloh[2], Mia Rizkinia[1], Ardiansyah[2], Rahmat Arief[1], Bambang Setiadi[3], Luhur Bayuaji[3], Josaphat Tetuko Sri Sumantyo[3]

[1]Department of Electrical Engineering, Faculty of Enginering, Universitas Indonesia, Kampus Baru UI, Depok 16424, Indonesia.
[2]Department Geography, Faculty of Mathematics & Natural Sciences, Universitas Indonesia, Kampus Baru UI, Depok 16424, Indonesia.
[3]Josaphat Microwave Remote Sensing Laboratory, Center for Environmental Remote Sensing, Chiba University, 1-33 Yayoi-cho, Inage-ku, Chiba 263-8522, Japan.

Email: dodi.sudiana@ui.ac.id

Abstract. The geographical position of Indonesia located between two continents and oceans is strategic, but at large risk of experiencing various disasters. Climate change and vulnerable location (surrounded by plates and geological faults in the Earth's crust) creates an earthquake-prone region and causes land/mudslides. In this paper, PS InSAR method (Persistent Scatterer Interferometric Synthetic Aperture Radar) is implemented to Phased Array type L-band Synthetic Aperture Radar (PALSAR) data to study the potential damage caused by the earthquake or volcanic eruption in Malang vicinity. By comparing the amplitude images periodically, shifting soil can be determined using precise orbital information. The analysis showed a significant decrease of land deformation on slope area in Klojen district in Malang city, reached up to –7.128 mm/year.

1. Introduction

Indonesia is one of countries which is prone to natural disasters. In some area, e.g. Bandung, Jakarta, Medan, Malang, Semarang and other cities, land deformations, land subsidence and landslide are main cause of building cracks, floods, changing land use, and other serious problems. Synthetic Aperture Radar (SAR) data processing using Differential Interferometric Synthetic Aperture Radar (DInSAR) has proven to be the effective solution to measure land deformation precisely in wide spatial coverage with high resolution data. Many applications are implemented using this method, such as land subsidence in Bandung [10], Jakarta [11], Sidoarjo mudslide [12], glacier movement [20], volcanic activities [9][19], Earth's crust movement [7][21], and underground mining activities [13][3][4].

However, DInSAR method has a weakness such as prone to temporal changes and atmospheric disturbance. In the paper, we propose Persistent/Permanent Scatterer Interferometry (PSI) or PS InSAR method which is a newer method developed from DInSAR to analyze land deformation on slope area in Malang vicinity, East Java, Indonesia. This method is based on multi-temporal SAR imagery data to enhance the detection sensitivity for land deformation [1]. The backscattered signal from a coherent object during data acquisition, which are known as permanent/persistent scatterer (PS)

[4] To whom any correspondence should be addressed.

is analyzed. It is important to have these permanent pixels since using clear interferograms, these positions represent land movement. In SAR imageries, human made objects could be found clearly in cities or high slopes. The PS InSAR is developed by scientists to monitor land deformation in many applications, including land deformations in cities [10][18][6], infrastructure stability [14], seismic faults [15], volcano activities [2], and landslides in sloped area [17]. The advantages of this method are: (1) wide area coverage , (2) efficient for mapping and survey activities, and (3) high accuracy in measuring land deformation as well as conventional land survey. Moreover, the method offers high scalability, effective since there is no in-situ measurements, time and cost efficient, almost real-time for systematic monitoring system, although it has high dependency on data provided by ALOS satellite and high computation complexity.

PALSAR is a radar sensor carried by ALOS satellite. Since it works in microwave band, comparing to optical sensor, it becomes reliable sensor working in any weather conditions, day and night. During normal operation, PALSAR observes the Earth's surface with 34.3° off-nadir angle to produce 10 meter spatial resolution mode. Off-nadir angle could be changed from 9.7° to 50.8°. Wide observation mode could be achieved using ScanSAR which has 100 meter spatial resolution. Swath width mode of ScanSAR is from 250 to 350 km, while normal resolution has 70 km wide.

There are three level of PALSAR product data:

1. Level 1.0
   Data which is reconstructed from raw data, unprocessed signal with additional geometric and radiometric coefficients correction, but not used yet. Imageries are separated for each polarization (HH, VV, HV, VH).
2. Level 1.1
   In this level, data has been compressed using range compression and azimuth compression and stored as complex data for slant range. Each polarization separates each imagery.
3. Level 1.5
   Imageries are processed in multi-look and projected into a geographical position. Longitude and latitude in the product are calculated without considering elevation using systematic geocoding (G) or systematic georeference (R). Each imagery is separated based on the polarization and only one technique (G or R) is applied.

Basically, PALSAR data acquisition is the same as SAR acquisition mode. Based on configuration system, SAR sensor could retrieve the data in several modes: Stripmap, ScanSAR dan Spotlight.

2. Persistent Scatterer Interferometry SAR (PS InSAR)

PS InSAR is developed from InSAR (Interferometric Synthetic Aperture Radar), which is used in 90s to measure land deformation. The objective of PS InSAR application in the beginning is to identify land deformation in single coherent pixels of several SAR imageries separated by large baseline in order to achieve high accuracy. The level of land deformation accuracy at least up to sub-meter fraction or even millimeters per year. Since DInSAR utilizes multitemporal radar imageries, temporal decorrelation and atmospheric inhomogeneous degrade the interferogram. Other disadvantages are: (1) geometric decorrelation related to the large baseline between two acquisition imageries, (2) the ambiguity that limits the operational power of this method. Permanent Scatterer (PS) could eliminate the effects and increase the land deformation accuracy. The PS InSAR process is started by identifying all PS points and analyze the 2D deformation in these points which could eliminate main weaknesses of DInSAR. This is based on the fact that the atmospheric phase contribution correlated spatially to single scene of SAR data, but tend to uncorrelated in daily or weekly data. Therefore, the atmospheric effect could be estimated and eliminated by combining long period data analysis and temporal fluctuation could be calculated [13]. The process of PS InSAR is briefly described in Fig. 1 [16][1]. Key steps of the process are: (1) interferogram computation, (2) differential interferogram computation using Digital Elevation Model (DEM), (3) preliminary estimation of the most possible coherent pixels, (4) improving step (3), where rough grid of PS candidates estimate the wavelength which represents the atmospheric signal. After interpolate these estimations, differential interferogram is corrected and additional PS are recalculated [9].

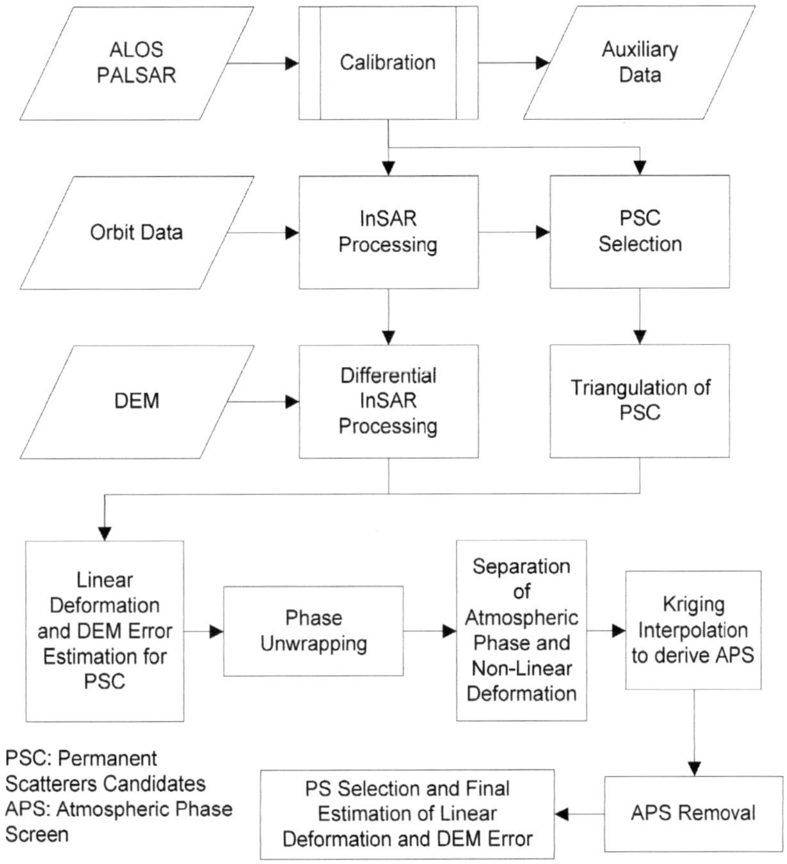

Figure 1. Block diagram of PS InSAR processing.

3. Area and Data Description

Malang is a city which is located at an altitude between 440-667 meters above sea level, and one of tourist destinations in East Java because of its natural and climatic beauty. It is located in the midst of Malang regency from 7.06°-8.02°S latitude to 112.06°-112.07°E longitude. The regency is surrounded by mountains: Mt. Arjuno in the North, Mt. Semeru in the East, Mt. Kelud in the south, Mt. Kawi and Panderman in the West. Land condition in Malang, among others: the southern part includes a wide plateau (suitable for industry), the northern part of the plateau is fertile (suitable for agriculture), eastern part is plateau with less state less fertile, the western part is a vast plateau (area suitable for education).

To analyze the land deformation in Malang and its vicinity, 6 (six) ALOS/PALSAR data in SLC format are used. To eliminate the topographical phase, low resolution DEM data from Shuttle Radar Topography Mission (SRTM) in 90×90m resolution is used as external DEM. PALSAR data are retrieved from February 2010 to February 2011 period and among those data, the oldest is selected as master image to calculate the temporal interferogram. Interferometric combination is derived to assure uniform spatial and temporal baseline distribution.

4. Experiment Results

The study area is limited to the district of Klojen, Malang city. The geographical location is 7.953-7.991°S and 112.617-112.639° E. The analysis result is land deformation of the district and presented as vector file for the boundary of 107.552°E to 107.732°E, -6.9749°S to -6.8787°S. The deformation is overlaid on basic map derived from Google Map as shown in Fig. 2.

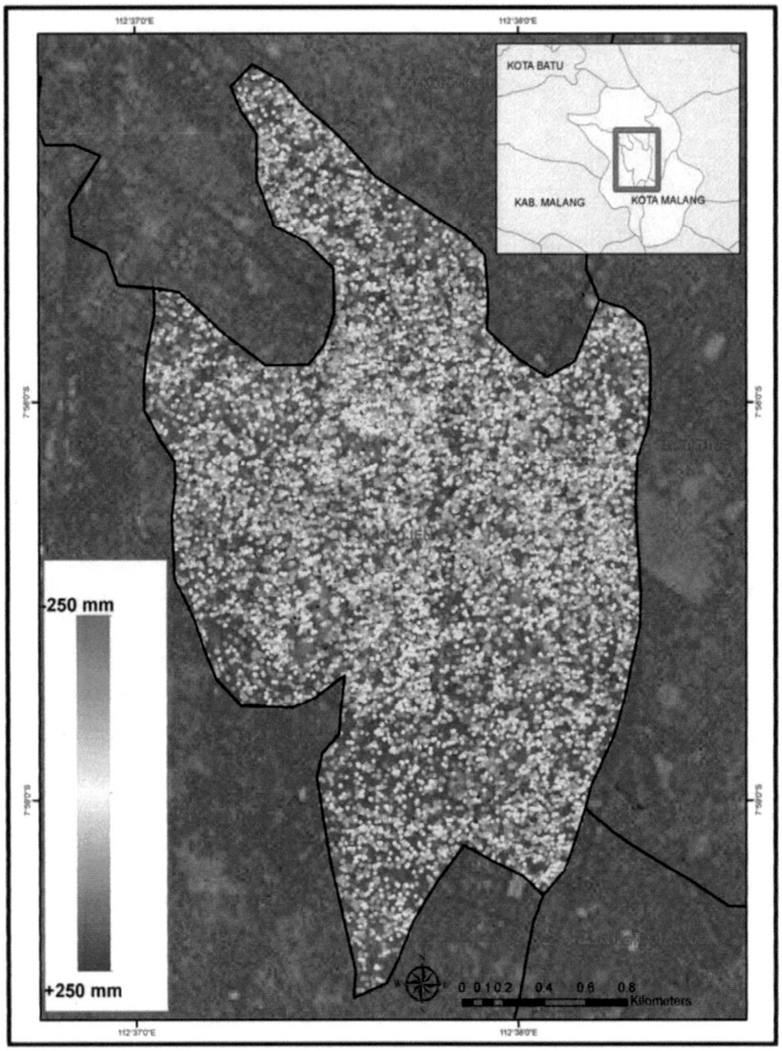

Figure 2. Velocity map of land deformation in Klojen District, Malang city

The district is divided into 11 villages, namely Bareng, Gadingkasri, Kasin, Kauman, Kidul Dalem, Klojen, Oro-oro Dowo, Penangungan, Rampal, Samaan, and Sukoharjo. The oldest PALSAR data retrieved on February 2010 is selected as master file. The highest averaged displacement occurs in Bareng village from 2010-2011, reaches -7.128 mm/year. And the lowest is Penangunan village which has 0.501 mm/year. Deformation velocity of these villages is presented in Table 1.

Table 1. Deformation time series for 11villages in Klojen district, Malang city

Village	Master	Slave					Averaged Displacement (mm/year)
	Feb-10	Jul-10	Oct-10	Nov-10	Jan-11	Feb-11	
Bareng	0.000	2.070	0.507	-2.268	-1.750	-5.688	-7.128
Gadingkasri	0.000	3.518	2.675	0.124	-0.360	-1.965	3.992
Kasin	0.000	2.902	2.443	-1.499	-1.615	-4.483	-2.253
Kauman	0.000	0.292	-0.265	-0.356	0.037	-2.795	-3.086
Kidul Dalem	0.000	-0.130	-2.107	0.425	1.230	-1.553	-2.135
Klojen	0.000	-1.205	-2.270	-0.588	-0.883	-0.888	-5.834
Oro-oro Dowo	0.000	0.435	-0.585	0.618	1.092	-0.453	1.107
Penangungan	0.000	-2.045	-2.898	1.428	2.051	1.964	0.501
Rampal	0.000	-1.551	-1.603	2.082	2.771	3.622	5.322
Samaan	0.000	-1.270	-1.128	0.954	0.808	2.050	1.414
Sukoharjo	0.000	0.605	0.413	-1.957	-1.857	-3.892	-6.688

5. Analysis

Displacement located at Bareng village, has a significant character which has gradually increasing displacement, starting from 2.070 mm/year up to -7.128 mm/year, as shown in Fig. 3. Bareng is one of main village in Klojen district in Malang city, where trade area, tourist area and buildings are located. Eight other villages have fluctuating displacement values as shown in the figure. But it could be concluded that the displacement trend between October 2010 to February 2011 is positive for Kasin, Penangunan, Samaan villages, and negative trends for the rest of villages.

Malang consists of alluvial sediment material at 59% of the total area [5]. It becomes one factor of displacement in this area which alluvial sediment gives contribution of the displacement because of its characteristic. In addition, 10% area of Malang consists of Mediterranean lands which classified as erosion-high sensitivity, 15% of Andosol which classified as erosion-very high sensitivity, and 16% of Latosol as erosion-medium sensitivity.

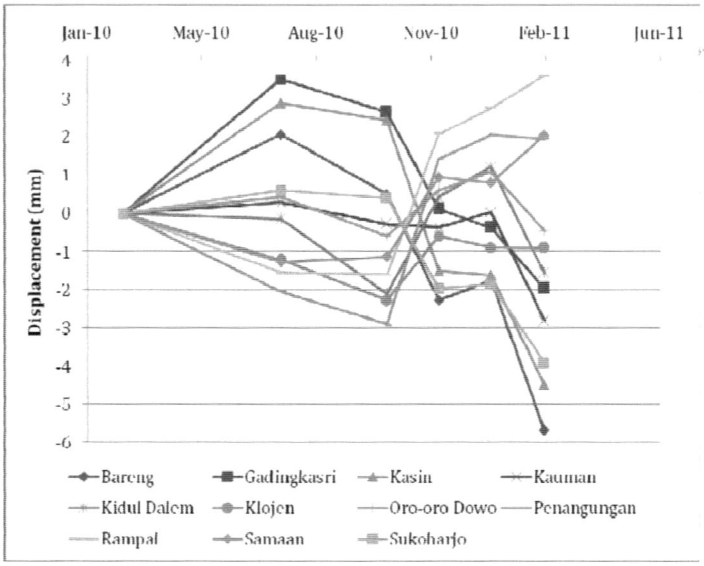

Figure 3. Displacement of 11 villages in Klojen district, Malang city using PS InSAR method

6. Conclusions

It has been demonstrated that PS InSAR technique is applicable for mm scale surface deformation monitoring in Malang. By analyzing data set of ALOS/PALSAR images covering the city of Malang, significant subsidence of up to –7.128 mm/year has been detected between year 2010-2011. To analyze longer deformation phenomena, longer period of PALSAR data is needed. The deformation estimates relate well with results from an independent study and geodetic measurements.

References

[1] Ferretti A, Prati C, and Rocca F 2001 *IEEE Trans. on Geosci. and Rem. Sens.* 39 **1** 8-20
[2] Hooper A J 2006, *Persistent Scatterer Radar Interferometry for Crustal Deformation Studies and Modeling of Volcanic Deformation* (Stanford Univ.)
[3] Ng AH-M, Chang C, Ge L, Rizos C, and Omura M 2009 *Earth, Planets and Space* 61 **6** 733-745
[4] Ng AH-M, Ge L, Yan Y, Li X, Chang H-C, Zhang K, and Rizos C 2010 *Eng. Geol.* **1151-2** 1-15
[5] Ajie B, Firman H, and Andeas D 2008 *Went-Ristek's FOSS GIS & Sahana*
[6] Osmanoglu B, Dixon T H, Wdowinski S, Cabral-Cano E, and Jiang Y 2011 *Intl. J. of Appl. Earth Obs. and Geoinf.* 13 **1** 1-12
[7] Zebker H A et al 1994 *J. of Geophys. Res.* 99 **B10** 19,617-634
[8] Pemerintah Kota Malang, http://www.malangkota.go.id/mlg_halaman.php?id=1606076.
[9] Abidin H Z, Djaja R, Andreas H, Gamal M, Hirose K and Maruyama Y 2004 *Geomat. Res. Aust.* **81** 45-58
[10] ITT Visual Information Solutions **23**
[11] Sumantyo J T S, Shimada M, Mathieu P, and Abidin H Z 2012 *IEEE Trans. on Geosci. and Rem. Sens.* 50 **1** 259-270
[12] JAXA/EORC *ALOS user handbook*, 2005
[13] Yu J, Ng A H-M, Jung S, Ge L, and Rizos C 2008 *Intl. Arch. of the Photogram., Rem. Sens. and Spatial Inf. Sci.* XXXVII B1
[14] Bayuaji L, Sumantyo J T S, and Kuze H 2010 *Canad. J. of Rem. Sens.* 36 **1** 1-8
[15] Bayuaji L, Sumantyo J T S, and Kuze H 2011 *Proc. IEEE Int. Geosci. and Rem. Sens. Symp.* (IGARSS) 4391-4394
[16] Jiang L and Lin H 2010 *Eng. Geol.* 110 **3-4** 77-92
[17] Ono M 2009 *Rem. Sens. Tech. Center of Japan*, Tech. Rep.
[18] Adam N, Kampes B, and Eineder M 2004 *Proc. of the Envisat & ERS Symposium*, Salzburg, Austria
[19] Farina P, Colombo D, Fumagalli A, Marks F, and Moretti S 2006 *Eng. Geol.* 88 **3-4** 200-217
[20] Lanari R and Sansosti E 2008 *IEEE Radar Conf.* 1-6
[21] Gonzalez R C and Woods R E 2002 *Digital Image Processing* (Prentice Hall, Inc.)

Generation of orthogonal curvilinear grids on the sphere surface based on Laplace-Beltrami equations

Fuqiang Lu[1], Zhiyao Song[1,2] and Xin Fang[1]

[1] Key Laboratory of Virtual Geographic Environment (Ministry of Education), Nanjing Normal University, Nanjing, Jiangsu Province, 210046, P. R. China

[2] AA(Jiangsu Provincial Key Laboratory for Numerical Simulation of Large Scale Complex Systems, School of Mathematical Sciences, Nanjing Normal University, Nanjing 210046, China)

E-mail: [1]zhiyaosong@sohu.com

Abstract. In this paper, Laplace-Beltrami equations are used to generate orthogonal curvilinear grids on the sphere for ocean models. In addition to overcoming the pole-problem, the grid configuration has quasi-uniform cell-size on the whole sphere. Some quantities such as the grid length along two directions, the angle deviation from orthogonality, the area of the cell to evaluate the quality of the grid, which demonstrate the grid produced is fit to be a model grid on which the finite difference method or finite volume method can be implemented for numerical simulating of global atmosphere and ocean dynamics on large scale.

1. Introduction

The global climate is getting warmer, causing an increasing threat from extreme weather. It's necessary to forecast the severe change of global climate in order to make possible preparations to reduce the destruction. All this depends on the well development of the global climate model, in which the most important two components are the global atmosphere and ocean model. So far, the spherical grids used can be classified into two types[1]: one type is the structured grid, which is characterized by a regular layout, and the number of first-order neighbors of each element is constant, such as the latitude-longitude spherical grid, cubed-sphere grid, the Yin-Yang overset grid. The other is unstructured grid, which is composed by triangles or irregular polygons, the neighbors of each element are not so clear, such as the sphere icosahedral grid or geodesic grid.

The orthogonal curvilinear grid belongs to a structured grid, it has two major merits: at first, the code of mesh and neighbor-searching is obvious, and the dynamic model equations have less terms related to mesh metrics under the orthogonal curvilinear coordinate. Secondly, the finite difference method or the finite volume method can be implemented with much easy.

At present, many operational weather and climate models are based on a latitude-longitude spherical grid, however the convergence of the meridians leads to resolution clustering at the poles, which causes the well-known pole-problem[2]. Meanwhile, some methods such as zonal filtering or space-smoothing, semi-implicit or Semi-Lagrangian method have been developed to overcome this problem in the numerical simulation of global atmosphere dynamics, however it is difficult to use these methods in the ocean models[3], therefore it's necessary to consider a new kind of orthogonal curvilinear spherical grid, which is fit for numerical modelling of large-scale atmosphere and ocean dynamics and expression of geographical space information.

2. Generation of quadrilateral grids on the sphere surface

2.1. Gnomonic projection

Consider the cube with sides of length $2a$ inscribed into a sphere of radius R such that the eight vertices of the cube exactly touch the sphere and $R = \sqrt{3}a$. The cube is oriented in such a way that the 3D absolute Cartesian coordinate axes (X, Y, Z) are normal to the faces, and the centroid of the cube is at the origin, as it can be seen in Figure 2.1, the left sub-figure gives the layout of each face of an open cube, the right one shows the cube and its circum-sphere under 3D absolute Cartesian coordinates. Let (x, y) be the local Cartesian coordinates centered on the surface of the cube face such that $-a \le x, y \le a$ and (λ, φ) denotes spherical longitude-latitude coordinates, it follows that:

$$\begin{cases} X = R\cos(\lambda)\cos(\varphi) \\ Y = R\sin(\lambda)\cos(\varphi) \\ Z = R\sin(\varphi) \end{cases}$$

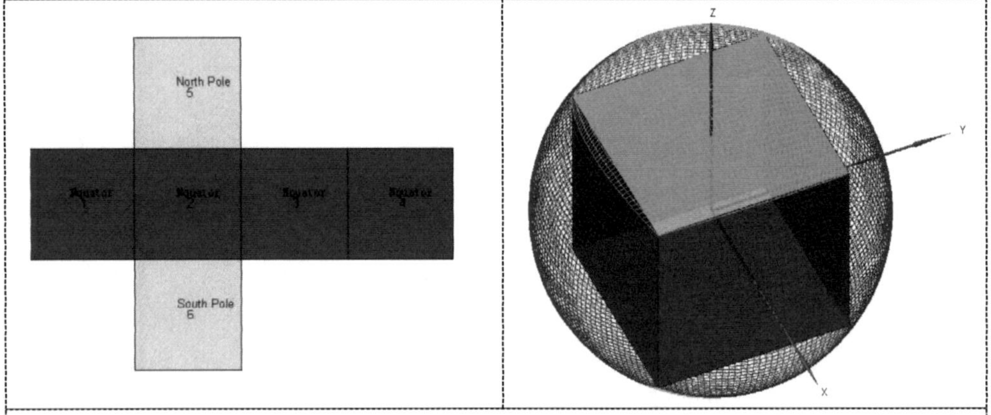

Figure 2.1. The left sub-figure gives the layout of an open cube, the right one shows the position of the cube and its circum-sphere under 3D absolute Cartesian coordinates.

There are two kinds of methods to generate spherical quadrilateral grids[4]

- Equidistant gnomonic mapping:

The relation between the 3D Cartesian coordinates on the sphere and the local 2D Cartesian coordinate on cube surface $X = a$, see figure 2.2, which shows the relation between a point (x, y) in local 2D Cartesian coordinates and a point (X, Y, Z) on the sphere with radius R:

$$(X,Y,Z) = \frac{R}{r}(a, x, y)$$

where r is the Euclidean distance between (a, x, y) and the origin.

$$r = \sqrt{a^2 + x^2 + y^2}$$

- Equiangular gnomonic mapping:

$$\begin{cases} x = a * \tan(\mathrm{u}) \\ y = a * \tan(\mathrm{v}) \end{cases}$$

it is obvious that $u, v \in [-\frac{\pi}{4}, \frac{\pi}{4}]$.

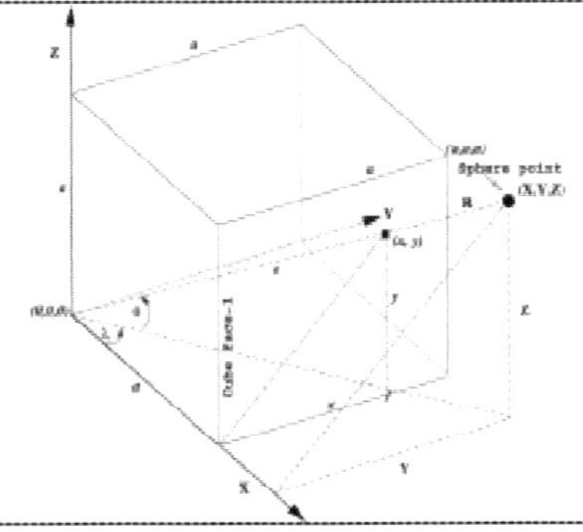

Figure 2.2. Gnomonic projection of the cube surface $X = a$ on the sphere

Rančic' et al[5] showed that the equiangular projection produces a more uniformly spaced grid compared to the equidistant projection and is more suitable for finite difference method. D.Nair[4] indicated that on the sphere, with the same number of grid points, an equiangular projection is more accurate than an equidistant projection. But these two grids are not orthogonal and dynamic equations would become more complex under these non-orthogonal curvilinear coordinates.

2.2. Elliptic grid generation on surface

Elliptic grid generation is the iterative relaxation of a first-guess grid (via successive over relaxation) to satisfy a quasi-linear elliptic system of partial differential equations (PDEs) while imposing smoothness and orthogonality constraints[6]. In the case of the grid on sphere surface, each face of the physical grid in spherical coordinates is mapped to a 2-dimensional rectangular u-v parametric space. The elliptic systems are solved in parametric space while imposing the desired physical constraints; orthogonality and uniform grid spacing are the most important constraints in most of practical application. There are two options for imposing orthogonality constraints: the Neumann constraint that allows the grid points to slide along the boundary of each face, and the Dirichlet constraint that holds the edge points fixed while adjusting the interior points and maintaining a quasi-uniform grid cell spacing. The Neumann approach computes a shift for each grid point along the boundary of the grid by forcing the dot product of the physical variables on two orthogonal coordinates (ξ, η) partial derivatives to be equal to zero; this approach will produce a convergence of grid cells near the corners. Orthogonality may also be imposed through the use of orthogonal control functions, i.e. the Dirichlet approach. These control functions are evaluated at the boundary to impose orthogonality, while maintaining the original grid spacing along the boundaries. The computation of two-order normal derivatives in control functions depends on a layer of ghost boundary points, getting the positions of these ghost points is the key. Then transfinite interpolation can be applied to interpolate the control functions from the boundary to the interior of the grid. In addition, a blending function can also be applied to force the interior points to remain close to the original grid spacing.

In classical differential geometry, a surface S is viewed as a mapping from R^2 to R^3.Consequently, parametric surfaces or physical variables are defined in terms of parametric variables. In grid generation, parametric variables are defined in terms of computational variables, that is:

$$\overrightarrow{X} = (x, y, z) = (x(u,v), y(u,v), z(u,v)) \quad (1)$$

$$(u,v) = (u(\xi,\eta), v(\xi,\eta))$$

In many cases, the conformal mapping can be used to generate the Body-Fitted mesh on a simple-connected domain under 2D cartesian coordinates. Similarly, a conforming mapping of a smooth bounded surface onto a rectangular region can be constructed by establishing a mapping from a square region of the computational plane into the surface which is orthogonal and has a constant aspect ratio[7,8]. The conditions can be expressed by the system of equations:

$$\begin{aligned} \overrightarrow{X}_{\xi} \cdot \overrightarrow{X}_{\eta} &= 0 \\ M\left|\overrightarrow{X}_{\xi}\right| &= \left|\overrightarrow{X}_{\eta}\right| \end{aligned} \quad (2)$$

where M is the grid aspect ratio. These two equations can be rewritten as:

$$X_{\xi}X_{\eta} + Y_{\xi}Y_{\eta} + Z_{\xi}Z_{\eta} = 0 \quad (3)$$

$$M^2\left(X_{\xi}^2 + Y_{\xi}^2 + Z_{\xi}^2\right) = X_{\eta}^2 + Y_{\eta}^2 + Z_{\eta}^2 \quad (4)$$

Using the chain rule for differentiation, the above equations can be expressed in the form:

$$\begin{aligned} Mu_{\xi} &= av_{\eta} - bu_{\eta} \\ Mv_{\xi} &= bv_{\eta} - cu_{\eta} \end{aligned} \quad (5)$$

Where

$$a = \frac{\bar{g}_{22}}{\bar{J}}, b = -\frac{\bar{g}_{12}}{\bar{J}}, c = \frac{\bar{g}_{11}}{\bar{J}}$$

$$\bar{g}_{11} = (x_u)^2 + (y_u)^2 + (z_u)^2 \tag{6}$$

$$\bar{g}_{12} = x_u x_v + y_u y_v + z_u z_v$$

$$\bar{g}_{22} = (x_v)^2 + (y_v)^2 + (z_v)^2$$

$$\bar{J} = \sqrt{\bar{g}_{11}\bar{g}_{22} - (\bar{g}_{12})^2}$$

Assuming the relation between the computational variables ξ, η and the parametric variables u, v is:

$$\begin{cases} \xi = \xi(\mathrm{u,v}) \\ \eta = \eta(\mathrm{u,v}) \end{cases}$$

let ξ and η are twice continuously differentiable and the Jacobian of the inverse transformation $J = u_\xi v_\eta - u_\eta v_\xi$ is non-vanishing in the region under consideration. Then the metrics $u_\xi, u_\eta, v_\xi, v_\eta$ and $\xi_u, \xi_v, \eta_u, \eta_v$ are uniquely related by

$$\begin{aligned} \xi_u &= \frac{v_\eta}{J}, \xi_v = -\frac{u_\eta}{J} \\ \eta_u &= -\frac{v_\xi}{J}, \eta_v = \frac{u_\xi}{J} \end{aligned} \tag{7}$$

Using these quantities in equation (5) so that the parametric variables become the independent variables, the system can be expressed in the form

$$\begin{aligned} M\eta_v &= a\xi_u + b\xi_v \\ -M\eta_v &= b\xi_u + c\xi_v \end{aligned} \tag{8}$$

solving the above systems for ξ_v, ξ_u

$$\begin{aligned} \xi_\mathrm{v} &= M(c\eta_v + b\eta_u) \\ -\xi_\mathrm{u} &= M(b\eta_u + a\eta_u) \end{aligned} \tag{9}$$

note that $b^2 = ac - 1$ is used. This first-order system is analogous to Beltrami's system of equations for the quasi-conformal mapping on planar regions. It follows that the computational variables ξ and η are solutions of the following second-order linear elliptic system with $\Phi = \Psi = 0$

$$\bar{g}_{22}\xi_{uu} - 2\bar{g}_{12}\xi_{uv} + \bar{g}_{11}\xi_{vv} + (\Delta_2 u)\xi_u + (\Delta_2 v)\xi_v = \bar{J}\Phi \tag{10}$$

$$\bar{g}_{22}\eta_{uu} - 2\bar{g}_{12}\eta_{uv} + \bar{g}_{11}\eta_{vv} + (\Delta_2 u)\eta_u + (\Delta_2 v)\eta_v = \bar{J}\Psi \tag{11}$$

The Beltramians $\Delta_2 u$ and $\Delta_2 v$ are given by:

$$\begin{aligned} \Delta_2 u &= \bar{J}(a_u + b_v) = \bar{J}\left[\frac{\partial}{\partial u}\left(\frac{\bar{g}_{22}}{\bar{J}}\right) - \frac{\partial}{\partial v}\left(\frac{\bar{g}_{12}}{\bar{J}}\right)\right] \\ \Delta_2 v &= \bar{J}(b_u + c_v) = \bar{J}\left[\frac{\partial}{\partial v}\left(\frac{\bar{g}_{11}}{\bar{J}}\right) - \frac{\partial}{\partial u}\left(\frac{\bar{g}_{12}}{\bar{J}}\right)\right] \end{aligned} \tag{12}$$

The above system is the basis for elliptic methods generating surface grids. The source terms (or control functions) Φ and Ψ are added to allow control over the distribution of the grid points on the surface. Typically, the points in the computational space are given and the points in the parametric space must be computed. Therefore, it is convenient to interchange variables so that the computational variables ξ and η are the independent variables. The transformation given by equations (10) and (11) is reduced to the following system of equations for which the parametric variables u and v are the solutions of the following quasi-linear elliptic system[8,9]:

$$g_{22}(u_{\xi\xi} + Pu_\xi) - 2g_{12}u_{\xi\eta} + g_{11}(u_{\eta\eta} + Qu_\eta) = J^2\Delta_2 u \tag{13}$$

$$g_{11}(v_{\xi\xi} + Pv_\xi) - 2g_{12}v_{\xi\eta} + g_{11}(v_{\eta\eta} + Qv_\eta) = J^2\Delta_2 v \tag{14}$$

the above system is the so-called Lapalce-Beltrami equations, where

$$\begin{aligned} g_{11} &= \bar{g}_{11}(u_\xi)^2 + 2\bar{g}_{12}u_\xi v_\xi + \bar{g}_{22}(v_\xi)^2 \\ g_{12} &= \bar{g}_{11}u_\xi u_\eta + \bar{g}_{12}(u_\xi v_\eta + u_\eta v_\xi) + \bar{g}_{22}v_\xi v_\eta \\ g_{22} &= \bar{g}_{11}(u_\eta)^2 + 2\bar{g}_{12}u_\eta v_\eta + \bar{g}_{22}(v_\eta)^2 \\ P &= \frac{\bar{J}J^2}{g_{22}}\Phi, Q = \frac{\bar{J}J^2}{g_{11}}\Psi \\ J &= u_\xi v_\eta - u_\eta v_\xi \end{aligned}$$

In addition, the following equations could be computed beforehand to get the Beltramians $\Delta_2 u$ and $\Delta_2 v$

$$\left(\overline{J}\right)_u = \frac{1}{2\overline{J}}\left[\overline{g}_{11}\left(\overline{g}_{22}\right)_u + \overline{g}_{22}\left(\overline{g}_{11}\right)_u - 2\overline{g}_{12}\left(\overline{g}_{12}\right)_u\right]$$

$$\left(\overline{J}\right)_v = \frac{1}{2\overline{J}}\left[\overline{g}_{11}\left(\overline{g}_{22}\right)_v + \overline{g}_{22}\left(\overline{g}_{11}\right)_v - 2\overline{g}_{12}\left(\overline{g}_{12}\right)_v\right]$$

$$\begin{aligned}
\left(\overline{g}_{11}\right)_u &= 2\overrightarrow{X}_u \cdot \overrightarrow{X}_{uu} \\
\left(\overline{g}_{11}\right)_v &= 2\overrightarrow{X}_u \cdot \overrightarrow{X}_{uv} \\
\left(\overline{g}_{22}\right)_u &= 2\overrightarrow{X}_v \cdot \overrightarrow{X}_{uv} \\
\left(\overline{g}_{22}\right)_v &= 2\overrightarrow{X}_v \cdot \overrightarrow{X}_{vv} \\
\left(\overline{g}_{12}\right)_u &= \overrightarrow{X}_u \cdot \overrightarrow{X}_{uv} + \overrightarrow{X}_v \cdot \overrightarrow{X}_{uu} \\
\left(\overline{g}_{12}\right)_v &= \overrightarrow{X}_u \cdot \overrightarrow{X}_{vv} + \overrightarrow{X}_v \cdot \overrightarrow{X}_{uv}
\end{aligned} \tag{15}$$

we take the finite difference method to solve the equations (13) and (14), and the terms $\overline{g}_{11}, \overline{g}_{12}, \overline{g}_{22}$ can be computed according to the expression of the parametric surface beforehand[10]. In terms of the grid points in the interior of the region, the central difference is used to approximate the first-order and second-order derivatives:

$$\begin{aligned}
u_\xi &\approx \frac{u_{i+1,j} - u_{i-1,j}}{2\Delta\xi} \\
u_{\xi\xi} &\approx \frac{u_{i+1,j} - 2u_{i,j} + u_{i-1,j}}{(\Delta\xi)^2} \\
u_\eta &\approx \frac{u_{i,j-1} - u_{i,j-1}}{2\Delta\eta} \\
u_{\eta\eta} &\approx \frac{u_{i,j+1} - 2u_{i,j} + u_{i,j-1}}{(\Delta\eta)^2} \\
u_{\xi\eta} &\approx \frac{u_{i+1,j+1} - u_{i+1,j-1} - u_{i-1,j+1} + u_{i-1,j-1}}{4(\Delta\xi)(\Delta\eta)}
\end{aligned} \tag{16}$$

the partial derivatives of v can be approximated in the same way. Choose the discrete space-step $\Delta\xi = \Delta\eta = 1$, then

$$\begin{aligned}
u_{i,j} = \frac{1}{4\left(g_{22} + g_{11}\right)}\{ & g_{22}[2(u_{i-1,j} + u_{i+1,j}) + P(u_{i+1,j} - u_{i-1,j})] \\
& + g_{11}[2(u_{i,j-1} + u_{i,j+1}) + Q(u_{i,j+1} - u_{i,j-1})] \\
& - g_{12}(u_{i-1,j-1} - u_{i-1,j+1} - u_{i+1,j-1} + u_{i+1,j+1}) - 2J^2\Delta_2 u\}
\end{aligned} \tag{17}$$

$$v_{i,j} = \frac{1}{4(g_{22}+g_{11})}\{g_{22}[2(v_{i-1,j}+v_{i+1,j})+P(v_{i+1,j}-v_{i-1,j})]$$

$$+g_{11}[2(v_{i,j-1}+v_{i,j+1})+Q(v_{i,j+1}-v_{i,j-1})]$$

$$-g_{12}(v_{i-1,j-1}-v_{i-1,j+1}-v_{i+1,j-1}+v_{i+1,j+1})-2J^2\Delta_2 v\}$$

the above systems can be solved by the iterative method known as successive over-relaxation(SOR).

$$u_{i,j}^{n+1} = w_{i,j}u_{i,j}^{n+1} + (1-w_{i,j})u_{i,j}^{n}$$

$$v_{i,j}^{n+1} = w_{i,j}v_{i,j}^{n+1} + (1-w_{i,j})v_{i,j}^{n}$$

where the acceleration parameter $w_{i,j}$ should satisfy:

$$0< w_{i,j} <2$$

2.3. Neumann orthogonal boundary condition

We require the condition of orthogonality in physical space:

$$\overrightarrow{X}_{\xi} \cdot \overrightarrow{X}_{\eta} = 0 \quad \text{on} \quad \xi = 0,1,\cdots,m; \text{ and } \quad \eta = 0,1,\cdots,n \tag{18}$$

Where $\overrightarrow{X}$ is a composite function in equation (1), which takes on values in R^3. Expanding equation (2) using the chain rule yields the equation

$$\overline{g}_{11}u_{\xi}u_{\eta} + \overline{g}_{12}(u_{\xi}v_{\eta} + u_{\eta}v_{\xi}) + \overline{g}_{22}v_{\xi}v_{\eta} = 0 \tag{19}$$

This orthogonality condition is used to formulate derivative boundary conditions for the elliptic system. If the "left" and "right" boundary curves u = −1 and u = 1 are considered, we have $u_{\eta} = 0$ and the orthogonality condition reduces to

$$\overline{g}_{12}u_{\xi} + \overline{g}_{22}v_{\xi} = 0 \tag{20}$$

Similarly, along the "bottom" and "top" curves v = −1 and v = 1, $v_{\xi} = 0$ and orthogonality is imposed by

$$\overline{g}_{12}v_{\eta} + \overline{g}_{11}u_{\eta} = 0 \tag{21}$$

When solving the elliptic system, equation (20) determines the values of v on the boundary segments u = −1 and u = 1, and equation (21) determines the values of u on the boundary segments v = −1 and v = 1.

To implement this numerically, we use forward difference on the boundaries u = −1 and v = −1 and backward difference on the boundaries u = 1 and v = 1 to compute the new values for $u_{i,j}$ and $v_{i,j}$:

$$\begin{aligned}
v_{0,j} &= \frac{\overline{g}_{12}}{\overline{g}_{22}}(u_{1,j} - u_{0,j}) + v_{1,j} \\
v_{m,j} &= -\frac{\overline{g}_{12}}{\overline{g}_{22}}(u_{m,j} - u_{m-1,j}) + v_{m-1,j} \\
u_{i,0} &= \frac{\overline{g}_{12}}{\overline{g}_{11}}(v_{i,1} - v_{i,0}) + u_{i,1} \\
u_{i,n} &= -\frac{\overline{g}_{12}}{\overline{g}_{11}}(v_{i,n} - v_{i,n-1}) + u_{i,n-1}
\end{aligned} \tag{22}$$

2.4. The metrics of grid quality

We choose four major quantities used to analyze the grid quality: the grid length along ξ direction and η direction respectively (h_ξ and h_η), the cell-area of grid(area) and the angle deviation from orthogonality (DO), the ratio of h_ξ to h_η (aspect ratio).

$$\begin{aligned}
h_\xi &= \left\|\vec{X}_\xi\right\| = \sqrt{g_{11}} \\
h_\eta &= \left\|\vec{X}_\eta\right\| = \sqrt{g_{22}} \\
DO &= \left|\frac{\pi}{2} - \theta\right| \\
area &= \alpha_1 + \alpha_2 + \alpha_3 + \alpha_4 - 2\pi
\end{aligned} \tag{23}$$

where $\theta = \arccos\left(\frac{g_{12}}{\sqrt{g_{11}g_{22}}}\right)$, and $\alpha_i, i = 1,2,3,4$ is the spherical angle of each grid cell.

2.5. Numerical results

On the cube surface, the top panel $Z = a$ was chosen as a basis, on which a local 2D Cartesian coordinates (x, y) can be built, then stretch each points on the panel to its circum-sphere, i.e.

$$(X,Y,Z) = \frac{R}{r}(x, y, a)$$

where R is the radius of the sphere, $r = (x^2 + y^2 + a^2)^{\frac{1}{2}}$, $-a \le x, y \le a$.

Here we take $a = 1$, initial grid points distribution was chosen as equi-distance gnomonic cubed sphere grid, solving the elliptic equations (13) and (14) with Neumann orthogonal boundary condition equation (18), the convergence could be achieved after about 500 times iterations, the grid in parametric space with cell-number 44*44 was given in Figure 2.3, where the left sub-figure shows the initial grid points distribution in parametric space (u,v), the right one shows the final grid points distribution in parametric space after numerical iteration.

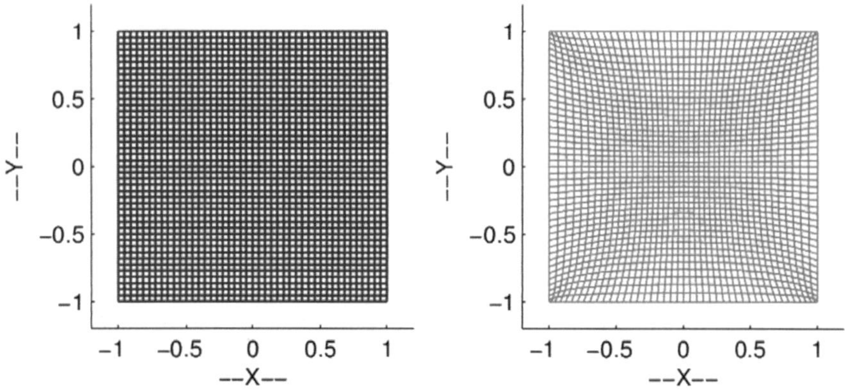

Figure 2.3. Initial parameter space in left sub-figure and final parameter space in right sub-figure with cell-number 44*44

At present, the cubed sphere grids could be also generated by other two methods, including equi-angular or equi-distance gnomonic projection and conforming mapping[5], moreover sphere grids produced by equi-angular or equi-distance gnomonic projection are not orthogonal, conforming mapping and elliptic cubed sphere grids are orthogonal. Comparison results about the grid metrics on cubed sphere grids between the three methods were given below.

We get the results under the condition that the grid cell-number is 22×22, 44×44, 90×90, 180×180 corresponding to approximate space resolution 4°, 2°, 1°, $30'$ respectively. Table 2.1 gives the maximum and minimum grid length, the maximum and average grid cell angle deviation from orthogonality excluding the four corners of the panel, the grid aspect ratio with different space resolutions on equi-angular gnomonic grids; Table 2.2, 2.3 give the same information but on grids produced by conformal mapping and elliptic-smoothing methods respectively.

Table 2.1. The grid quality on equi-angular cubed sphere grids with different cell-numbers.

cell-number	grid length		angle DO(degree)		aspect ratio	
	max	min	max	aver	max	aver
22×22	0.1238	0.0875	27.6229	8.0579	1.0000	1.0000
44×44	0.0620	0.0437	28.8149	8.0635	1.0000	1.0000
90×90	0.0304	0.0213	29.4215	8.0656	1.0000	1.0000
180×180	0.0153	0.0106	29.7109	8.0661	1.0000	1.0000

Table 2.2. The grid quality on conformal cubed sphere grids with different cell-numbers.

cell-number	grid length		angle DO(degree)		aspect ratio	
	max	min	max	aver	max	aver
22×22	0.1012	0.0436	12.8989	0.9250	1.0000	1.0000
44×44	0.0522	0.0181	12.9255	0.4625	1.0000	1.0000
90×90	0.0260	0.0071	12.9303	0.2261	1.0000	1.0000
180×180	0.0131	0.0029	12.9310	0.1131	1.0000	1.0000

Table 2.3. The grid quality on elliptic cubed sphere grids with different cell-numbers.

cell-number	grid length		angle DO(degree)		aspect ratio	
	max	min	max	aver	max	aver
22×22	0.1285	0.0299	12.9997	1.2770	1.0005	1.0000
44×44	0.0644	0.0116	12.7061	0.6744	1.0012	1.0000
90×90	0.0323	0.0043	12.5647	0.3500	1.0015	1.0000
180×180	0.0173	0.0018	12.5055	1.3875	1.0030	1.0000

Once the sphere grid on the top panel was produced, we get the whole sphere grids by proper solid-body rotation. Figure 2.4 shows the grid cell-angle deviation from orthogonality and the cell area variations on equi-angular sphere grids with cell number 44*44*6; while Figure 2.5, Figure 2.6 demonstrate the same information on the sphere grids generated by conformal mapping and elliptic equations iterations method respectively. From Figure 2.5, Figure 2.6, conformal mapping and elliptic-smoothing method produce the much like results, it can be seen that most of grid cell-angle deviation from orthogonality are not more than 10° with the exception of these in the neighborhood of the eight vertices. Even the conformal mapping method, the grids adjacent to the eight corners are also not exactly orthogonal. Under the same condition, the equi-angular gnomonic projection produced grids of the most uniform size, but are non-orthogonal, which was shown in Figure 2.4; while the conformal and elliptic-smoothing method generate quasi-orthogonal curvilinear grids on the cubed sphere at the cost of gridline weak-convergence on eight corner points.

IOP Conf. Series: Earth and Environmental Science **19** (2014) 012012 doi:10.1088/1755-1315/19/1/012012

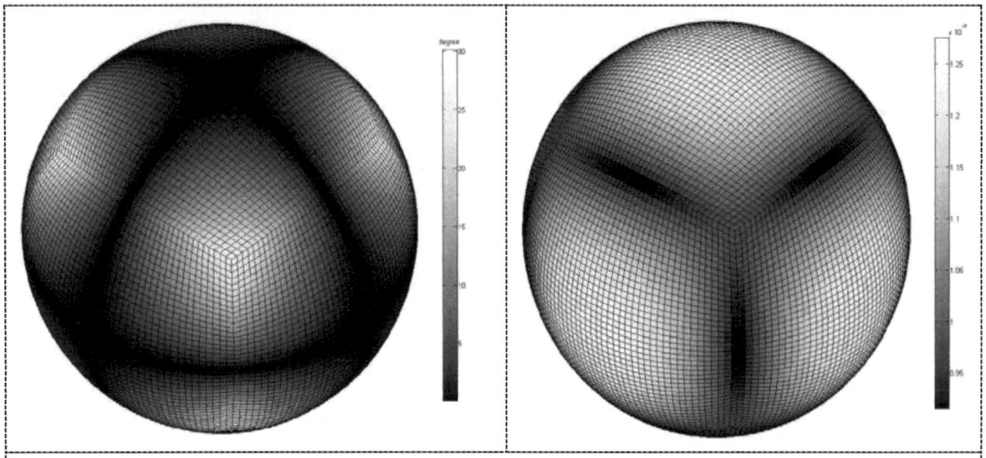

Figure 2.4. The left sub-figure shows the angle deviation orthogonality, the right one shows the cell area on equi-angular sphere grids with cell number 44*44*6.

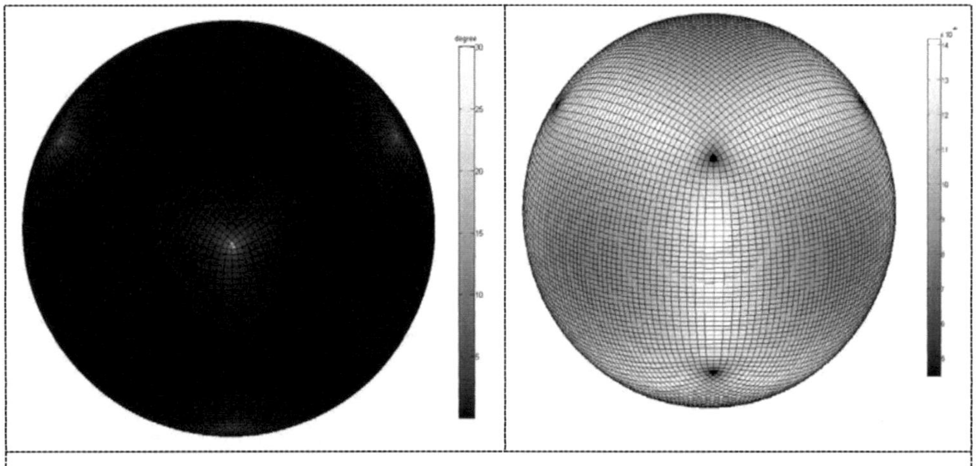

Figure 2.5. The same as above but on conformal mapping sphere grids with cell number 44*44*6.

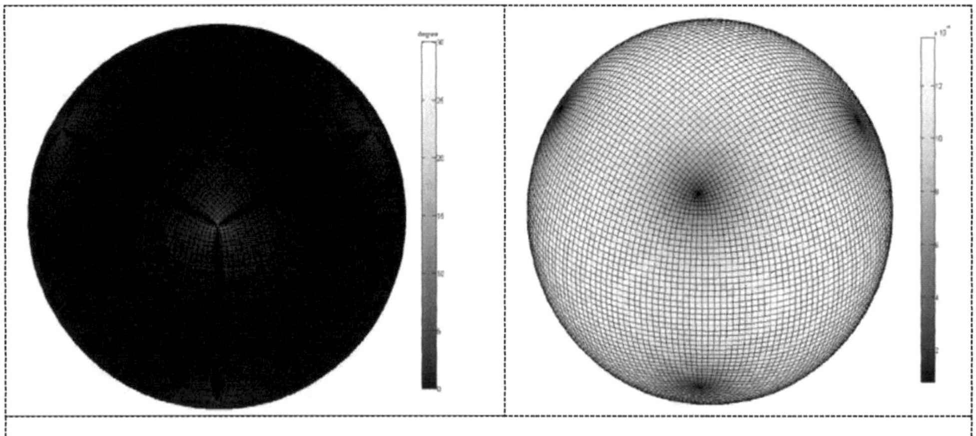

Figure 2.6. The same as above but on elliptic-smoothing sphere grids with cell number 44*44*6.

5. Conclusion

Basic properties of conformal mapping are utilized to improve methods of generating grids rom the solution of elliptic partial differential systems. Using the elliptic systems, A new kind of orthogonal curvilinear grid on the sphere was generated and the grid quality was evaluated according to four major quantities, including the grid length, cell-area, angle deviation from orthogonality and grid aspect ratio, moreover comparisons were made with the cubed sphere grids generated by equi-angular gnomonic projection and conformal mapping, the equi-angular gnomonic projection produced grids of the most uniform size, but are non-orthogonal; while the conformal and elliptic-smoothing method generate quasi -orthogonal curvilinear grids on the cubed sphere at the cost of gridline weak-convergence on eight corner points. This small defect would be eliminated if implicit method on time integration, which indicates that the orthogonal curvilinear spherical grid is satisfying wholly except for the neighborhood of eight vertices, so it can be chosen as a model grid on which the finite difference method or finite volume method can be implemented for numerical simulating of global atmosphere and ocean dynamics on large scale.

Acknowledgments
This project is partly supported by the State Key Program of National Natural Science of China (No.41231173), a grant from the National Natural Science foundation of China (grant No.41076008).

References

[1] Andrew Staniforth and John Thuburn. Horizontal grids for global weather and climate prediction models: a review. *Q. J. R. Meteorol. SOC.* **138**: 1-26, 2012.
[2] P.COURTIER and M. NAUGHTON. A pole problem in the reduced Gaussian grid. *Q. J. R. Meteorol.SOC.* **120**: 1389-1407, 1994.
[3] Alistair Adcroft, Jean-Michel Campin, Chris Hill, John Marshall. Implementation of an Atmosphere-Ocean General Circulation Model on the Expanded Spherical Cube. *Mon. Weather Rev.* **132**: 2845-2863, 2004.
[4] Ramachandran D Nair, Stephen J Thomas and Richard D. Loft. A discontinuous Galerkin Transport Scheme on the cubed sphere. *Mon. Weather Rev.* **133**: 814-828, 2004.
[5] M. Racic, R. J. Purser, and F. Messinger. A global shallow water model using an expanded spherical cube: Gnomonic versus conformal coordinates. *Q. J. R. Meteorol. SOC.* **122**: 959-982, 1996.
[6] William M. Putman, Shian-Jiann Lin. Finite-volume transport on various cubed-sphere grids. *J. Comput. Phys.* **227**: 55-78, 2007.
[7] S. P. S. Pekreijse. Elliptic grid generation based on laplace equations and Algebraic transformations. *J. Comput. Phys.* **118**: 38-61, 1995.
[8] C.W. MASTIN, J.F.THOMPSON. Quasi-conformal mappings and grid generation. *SIAM J.SCI. STAT.COMPUT.* **Vol.5**, No.2, June 1984.
[9] Ahmed Khamayseh, Bernd Hamann. Elliptic grid generation using NURBS surfaces. *Computer Aided Geometric Design* **13**: 369-386, 1996.
[10] Joe F. Thompson, BharatK. Soni, Nigel P. Weatherill. *Handbook of Grid Generation.* CRC Press, 1999, ISBN0-8493-2687-7.

Seismic Illumination Analysis in Poor Oil & Gas Field Data by Using Focal Beam Method

A H Abdul Latiff[1], D P Ghosh[2], Z Z Tuan Harith[2]

[1,2]Center for Seismic Imaging,
Universiti Teknologi PETRONAS,
Bandar Seri Iskandar, Ipoh,
31750 Malaysia

E-mail: abdulhalim.alatiff@petronas.com.my[1]

Abstract. The area underneath shallow gas cloud is an area where the image of subsurface data is generally poor. This distorted image underneath gas zones usually contains precious information of hydrocarbon accumulation. Previously, we analyse the factors contribute to poor subsurface seismic image underneath the gas cloud model and use focal beam technique to understand subsurface illumination information. Encourage by model-based success, we shift our focus to data-based application by applying the focal beam technique into a real field data. The results from this field were analyse in term of resolution function and amplitude versus ray parameter (AVP) imprint for different reflector depth, followed by acquisition analysis on the surface level. For this purpose, a velocity data of a field located in Malay Basin was built before applying the focal beam calculation. We will demonstrate that by using focal beam analysis for this field, we will able to obtain good imaging particularly for target reflector at 2000ms, 4000ms and 6000ms depth, provided the full 3D acquisition geometry was used during focal beam application.

1. Introduction

In seismically complex area such as shallow gas clouds and salt dome, acquiring and processing the seismic data using ordinary techniques normally will result in low illumination data. This is due to complex wave propagation behaviour, as severe wave distortion was detected from both scattering and attenuation phenomenon which resulted in blur subsurface seismic image, particularly underneath the gas accumulation. In order to restore the true earth cross section in seismic data, research had been focused in developing sophisticated imaging method, for example depth migration and reverse time migration, or by initiate some innovative way of survey design, like coil shooting technique and blended acquisition. This follows the forward path of seismic value chain, where a distinctive connection can be established between seismic acquisition and structural imaging. However, current imaging procedure require a lot of computational power and data space to determine viability of an acquisition design. With this constraint in mind, an integrated approach was develop in focal beam analysis [2] which enable faster evaluations of an acquisition design.

[1] Corresponding author.

The principle in focal beam is similar to downward continuation migration technique as both use shot and receiver profiles as their base. This method works by forward propagate shot profiles and cross-correlate with backward propagate receiver profiles. Once both profiles arrived on target point, an imaging condition will be applied to detect their similarity. Focal beam technique expand this idea as survey configuration was taken into consideration as well, to make the result more comprehensive. The beam will be produced by summing up all frequency content in a single wave propagation direction. Focal beam concept also relate to pre-stack migration formulation as both having double focusing process with the difference is in its output process. In pre-stack migration method, we need to simulate all shot records and before migrating the data while focal beam technique does not need to apply any prior processing algorithm. In addition, end user of focal beam method will get an insight of seismic data quality in shortest possible time while assess separately source geometry and receiver geometry configuration.

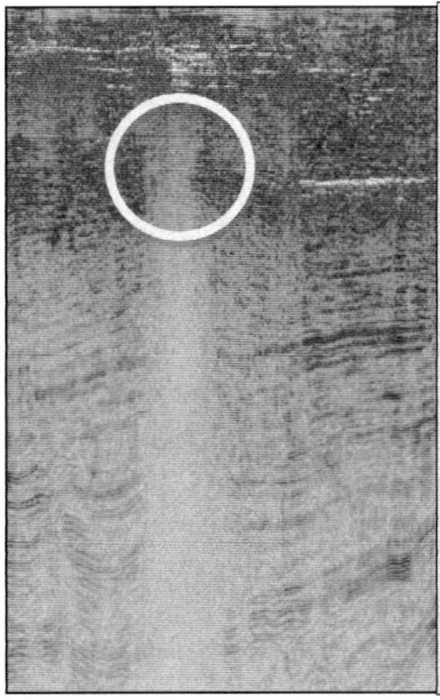

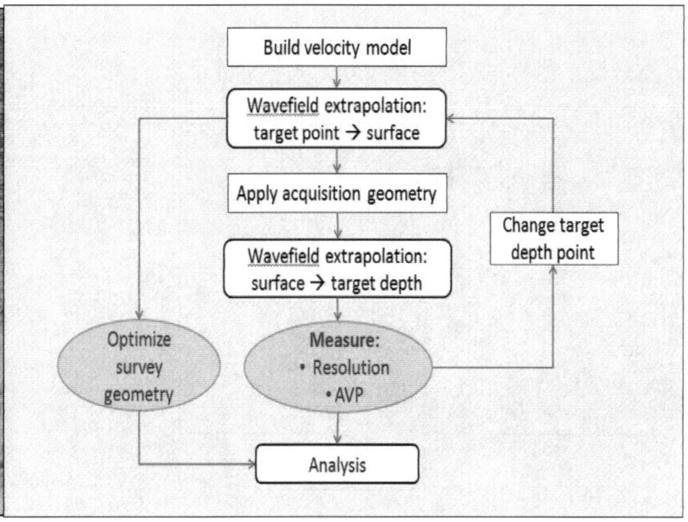

Figure 1 (Left): Theaffected gas cloud field in offshore Malaysia. It's clearly shown here the reflector underneath the gas cloud (circled) experience low illumination effect.
Figure 2 (Right): Workflow for application of focal beam analysis in gas cloud region.

The main importance of producing focal beam result is in term of active reservoir quantitative asseessment through subsurface structure information, designing acqusition geometry and choosing a migration operator. The workflow start by propagate each wave field frequency component from chosen target reflector towards surface position. On the surface, we design the desirable survey configuration, before the wave field being propagate back to target reflector. From both upward and downward wave, we analyze the result in spatial domain by multiplying both focal function, which producing a resolution function. Another option is to transform seismic data into radon domain, and analyze the result in amplitude versus ray-parameter function and this will indicate our illumination coverage of subsurface.

2. Methodology

In the shallow gas cloud affected area, a seismic image was affected in term of its kinematics (time sag) and dynamics (low amplitude). To restore its true reflectivity, the wave field propagation effects need to be removed by adding double focusing operators to remove downward and upward propagation. Continuation from beam concept, a procedure was develop which aim to integrated

IOP Conf. Series: Earth and Environmental Science **19** (2014) 012013 doi:10.1088/1755-1315/19/1/012013

seismic acqusition analysis into structural imaging of subsurface [3]. The steps being used in this study contain slight modification from our previous work [1], as now we will look into obtaining an suitable and practical acquisition configuration, and workflow in **Figure 2** summarise the sequence of our analysis.

The field velocity model used in the analysis (**Figure 3**) was built from smooth interval velocity file of 12.5m x 25m bin. The velocity measured for every 40 inline interval, each Common Depth Point (CDP) location and at 50ms time step. It means the distance between two consecutive velocity lines (y-direction) is 500m compared to 25m interval between two consecutive CDP locations (x-direction). Due to large gap in inline, it is important to interpolated y–direction velocity in our model, to prevent any discrepancies. However, due to limited computational power and resources, we limit our focal beam analysis into existing velocity points with modified bin of 20m x 20m. This give total dimensions for survey model approximately at 2.6km x 1km x 6 second.

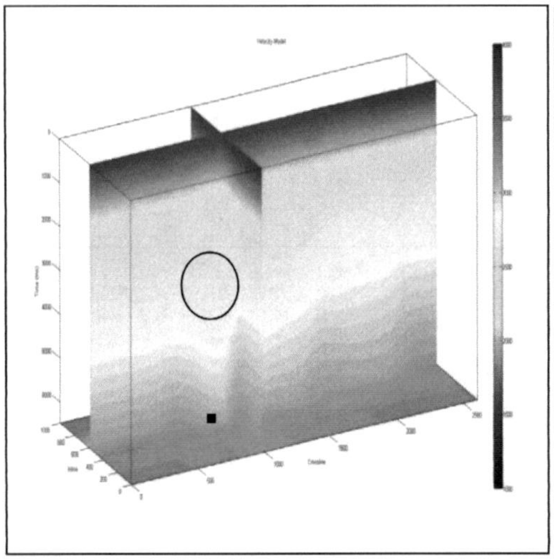
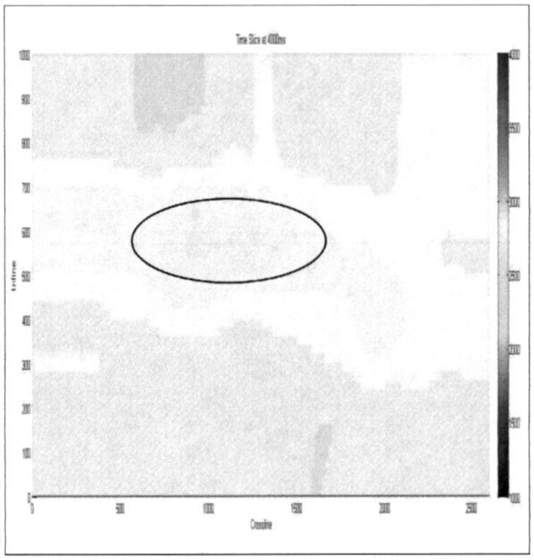

Figure 3 (Left): An inline slice and crossline slice of velocity model for a survey area in offshore Malaysia. The slices prove the existence of a shallow gas cloud field as the velocity value experience sudden 'pull-down effect' which indicate the low velocity region (circled). Black dot represent the lateral target point for illumination analysis.
Figure 4 (right): A time slice at time 4000ms depth was shown. Low velocity region indicate here in the circle.

For the purpose of investigation of illumination strength, three different depths of target points were chosen: depth at 2000ms, depth at 4000ms, and depth at 6000ms.These three points were chosen based on different reasons: 2000ms point is located just after the gas cloud region and provides clear idea how illumination behave in these zone. Meanwhile 4000ms point fall on a reflector, hence the illumination effect of the reflector can be understood thoroughly. The 6000ms point is at the end of recording data, means the significant comes in term of minimum illumination can be achieve in gas-affected data.The lateral location of the target points at each depth set just underneath the gas cloud area, i.e. [600, 1000] coordinate as indicate by black dot in **Figure 3**. It should be note that wave propagation at different depth level are independent of each other, means that there will be no interference between computation of 2000ms point reflector to 4000ms, and 6000ms point reflector and vice versa.

The gas cloud area fall in the region between 2000ms to 4000ms depths with velocity slightly deviated from surrounding sediments. In Figure 4, the area of gas cloud can easily identified as the velocity of those affected are much slower than similar depth sediments. Also it clearly shown in **Figure 4** that the accumulation of gas in shallower sediments has 'pull-down effect (black circle) which indicate the loss in its kinematic property. In order to investigate model's dynamic effect, forward modelling need to be simulated first before detail analysis can be carried out. Throughout focal beam computation, operator parameters used was chosen after consideration of earlier investigation and set constant for each depth point locations. The position of each target depth position will not affect other reflector point focal beam analysis. However, during the analysis we notice that more wave propagation was affected if the position is close to gas cloud due to bigger propagation angle during focal beam calculation. From the end result, we focus primarily on amplitude distribution and resolution function of the image compare to data quality as normally used in any commercial software.

For analysis purposes, we fix the acquisition to full 3D survey configuration during the focal beam calculation on the surface. Full 3D geometry has a unique characteristic where all sources and receivers will be located at the same coordinate and having the source / receiver lines spacing as minimal as possible. The advantage of having dense source-receiver network enable full wave field coverage on the surface level, while removes the possibility of missing out any upward wave. In the following design as shown in Figure 5, the geometry designed to follow the velocity model build up area of 2.6 km x 1 km with spacing of 20m in both x and y direction.

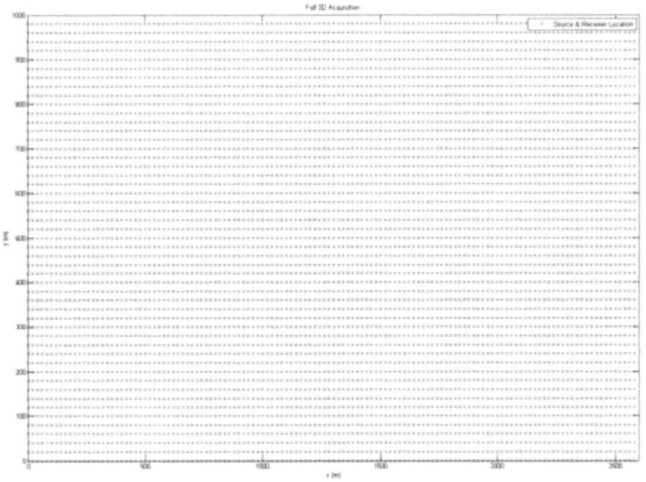

Figure 5: Top view of full 3D acquisition geometry used for illumination study.

3. Result & Discussion

3.1 Spatial Domain Analysis

Focal beam analysis was conducted after applying acquisition geometry to upward wave propagation, and propagate back to target reflector. Throughout the focal studies in both spatial and radon domains, the full 3D layout was used as the main objective is to find best achievable illumination of the target subsurface location. Note that in full 3D acquisition setup, source position is at the same place as receiver position, producing similar focal beam characteristic.

By multiplying both source and receiver beams for a monochrome frequency of 35Hz, the lateral resolution shown in **Figure 6** as it would be obtained for three different depth positions. Comparing focusing function obtained at different depths, the point which located closer to the surface has sharper resolution, whereas the response for target point at deeper locations is weaker due to waveform geometrical spreading effect. Note that throughout the analysis, attenuation factor and scattering effect has not been incorporated in the computation; hence the amplitude strength for deeper position is almost similar to shallower position.

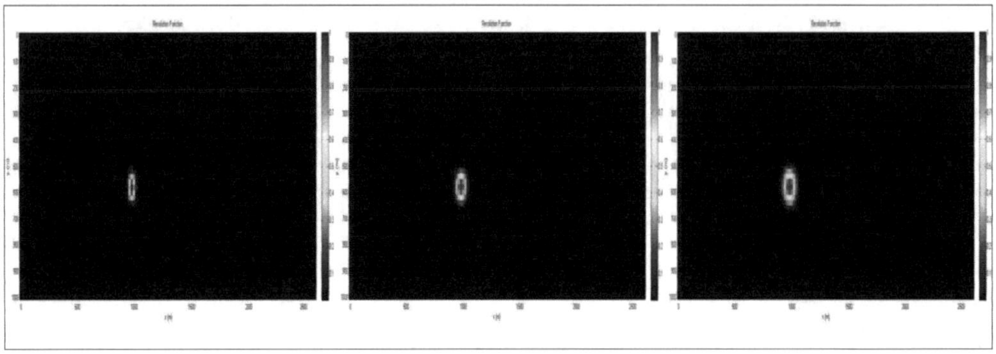

Figure 6: Resolution function for three different depth, at 2000ms (left), at 4000ms (middle) and at 6000ms (right). The resolution functions produce after multiplying focal source beam and focal receiver beam, which obtained at target reflector lateral position of [600, 1000]

3.2 Radon Domain Analysis

The product of focal source beam and focal receiver beam in radon domain will yield amplitude versus ray-parameter imprint (AVP). In **Figure 7**, AVP imprints for three target points at 2000ms, 4000ms and 6000ms has been analysed, repeating the procedure as in spatial domain. From the result, shallower position has wider illumination coverage due to fact less geometrical spreading occur compare to deeper position. However, all three AVP show common trait and similarities as wave propagation are less affected by deeper or shallower target locations. In addition, the weak velocity contrast between gas clouds to surrounding contributes to good AVP image of subsurface.

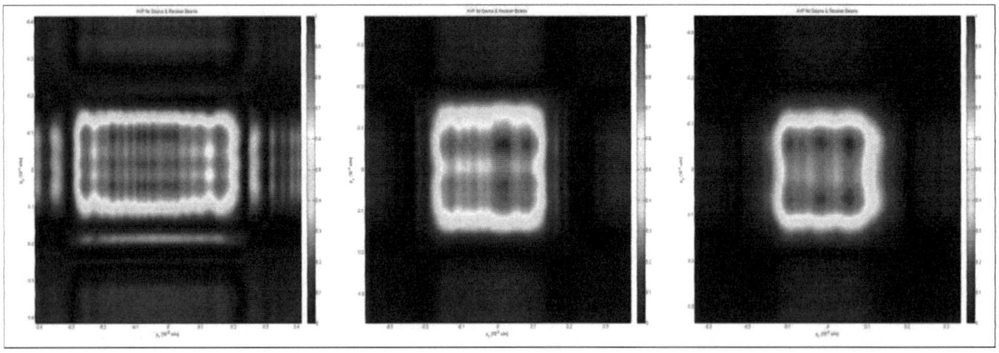

Figure 7: AVP function for three different depth, at 2000ms (left), at 4000ms (middle) and at 6000ms (right). Amplitude versus ray-parameter calculated by multiplying focal source beam and focal receiver beam in radon domain, which obtained at target reflector lateral position of [600, 1000]

4. Conclusion

In summary, the focal beam analysis in this particular field shows good imaging can be achieve particularly for our target points at 2000ms (located just underneath the gas cloud), 4000ms (falls on a target area of interest) and 6000ms (end of seismic data) depth, with justification of each chosen points explained in section two.However the good illumination coverage as shown in section three were obtained using the perfect acquisition geometry setup which not realistic given the constraint of time and resources of real worlds. Collectively, focal beam analysis in shallow gas cloud area can be summarised as follow:

1. Shallower depth point (2000ms point depth) will give better illumination coverage compare to deeper location (4000ms and 6000ms).
2. Weak gas cloud velocity is hardly affecting the illumination coverage of target subsurface due to less wave scattering and minimum attenuation occurred during wave propagation.
3. Full 3D acquisition produces high illumination coverage and good focusing for a particular survey area, albeit not possible in real practice.

Reference

[1] A.H.AbdulLatiff, D.P. Ghosh, Z.Z.TuanHarith, "Seismic illumination analysis of different shallow gas cloud velocities by focal beam method," Proceedings 2013 National Geoscience Conference, Ipoh, Malaysia,June 2013.

[2] E.J.VanVeldhuizen, "Integrated approach to 3D seismic acquisition geometry analysis," Delft University of Technology, PhD thesis, 2006.

[3] E.J.VanVeldhuizen, G.Blacquière, A.J.Berkhout, "Acquisition geometry analysis in complex 3D media," Geophysics, Vol. 73, p43–p58, 2008.

[4] A.J.Berkhout, L.Ongkiehong, A.W.F.Volker, G.Blacquière, "Comprehensive assessment of seismic acquisition geometries by focal beams—Part I: Theoretical considerations," Geophysics, Vol. 66, 2001.

[5] A.W.F.Volker, G.Blacquière, A.J.Berkhout, L.Ongkiehong, "Comprehensive assessment of seismic acquisition geometries by focal beams—Part II: Practical aspects and examples," Geophysics, Vol. 66, 2001.

IOP Conf. Series: Earth and Environmental Science **19** (2014) 012014 doi:10.1088/1755-1315/19/1/012014

Modelling animal waste pathogen transport from agricultural land to streams

Pramod K. Pandey[1], Michelle L. Soupir[2], Charles Ikenberry[2]

[1]Department of Population Health and Reproduction, University of California, Davis
[2]Department of Agricultural and Biosystems Engineering, Iowa State University, Ames, Iowa

E-mail:pkpandey@ucdavis.edu

Abstract. The transport of animal waste pathogens from crop land to streams can potentially elevate pathogen levels in stream water. Applying animal manure into crop land as fertilizers is a common practice in developing as well as in developed countries. Manure application into the crop land, however, can cause potential human health. To control pathogen levels in ambient water bodies such as streams, improving our understanding of pathogen transport at farm scale as well as at watershed scale is required. To understand the impacts of crop land receiving animal waste as fertilizers on stream's pathogen levels, here we investigate pathogen indicator transport at watershed scale. We exploited watershed scale hydrological model to estimate the transport of pathogens from the crop land to streams. Pathogen indicator levels (i.e., *E. coli* levels) in the stream water were predicted. With certain assumptions, model results are reasonable. This study can be used as guidelines for developing the models for calculating the impacts of crop land's animal manure on stream water.

1. Introduction

Bacterial pollution in stream water is a major concern [1]. For example, more than 50% of the assessed streams and rivers in the U.S. are impaired, and elevated pathogen/pathogen indicator levels are the leading cause of impairment [1]. The major sources of bacteria pollution in streams are agricultural activities. Agricultural non-point source pollution, particularly, animal waste in cropping land has a potential to elevate pathogen levels in stream water [2]. Excessive amount of manure produced in confined feeding operations (CAFOs) can potentially contaminate surface and ground water. As an example, more than 238,000 CAFOs in the U.S. produces more than 317 million gallons manure annually, managing such a huge quantity of manure can be a daunting task. Events such as rainfall after manure application in cropping land produces runoff, which can transport a large number of pathogens from farms to streams. In addition, events such as accidental spill from CAFOs can transport a large number of pathogens from CAFOs locations to streams [3, 4, 5, 6].

Mitigating the hazardous impacts of animal manure on environment is a serious issue. In addition to agricultural activities, the landscape characteristics of the watershed, soil, geology, land cover, and hydrology also play crucial roles in influencing stream water quality [7]. Predicting stream water quality requires understanding the various level of complexities and uncertainties involved in the watershed. One approach can be exploiting the hydrological models and Geographical Information

System (GIS) based dataset to understand the impacts of watershed characteristics (i.e., land cover, CAFOs, grazing, and feedlots) on the stream water quality. The GIS based data can provide the locational information about CAFOs, manure production, manure application rate, open feed lots, cropping land, and the land receiving animal manure; hydrological model can be used to understand the transport of pathogens from the watersheds via runoff to streams. The goal of this study is to utilize modelling approaches to understand the potential impacts of the watershed's animal waste on stream water quality.

There are many water quality models, which can be utilized for predicting stream water quality such as Better Assessment Science Integrating Point and Nonpoint Sources (BASINS), ArcView Generalized Watershed Loading Function (AVGWLF), Soil Water Assessment Tool (SWAT), and Watershed Analysis Risk Management Framework (WARMF). The models use watershed characteristics such as soil, land cover, elevation, and hydrology for predicting water quality. The BASINS was developed by the U. S. EPA to develop the Total Maximum Daily Load (TMDL) program, where the TMDL (mass per unit time) is the sum of the individual waste load allocations for non-point sources and natural background, load allocations for point sources, and a margin of safety. The AVGWLF was developed by Pennsylvania State University's Environmental Resources Research Institute for watershed assessment and TIMDL development. The SWAT model, which was developed by Texas A & M University, Black Land Research center, is a river basin scale model, which quantifies the influence of land management practices on large watershed and assesses the point and non-point pollution loading. These watershed scale models have been used extensively for estimating stream water quality such as the nutrient concentrations in stream water. Recently multiple studies have exploited the SWAT model for predicting stream water bacteria levels [8, 9, 10]. Considerable progresses are reported in improving stream water bacteria predictions utilizing the SWAT model. The objective of this paper is to describe briefly input data, model, processes, simulation approaches required for implementing the SWAT model and predicting stream bacteria levels. The details of the study are available elsewhere [10].

2. Study area descriptions

The work was conducted in the Squaw Creek Watershed, Iowa, USA. The study area and DEM are shown in Figure 1(top). Squaw Creek passes through four counties (Story, Webster, Hamilton, and Boone Counties), and it is a tributary of the South Skunk River. The total drainage area of the Squaw Creek watershed (Hydrological Unit Code (HUC 10)) is 592.39 sq km. The basin length and perimeter of HUC 10 watershed is 43.53 km and 134.02 km, respectively. The average slope of the watershed is 2.01% with the basin relief of 111.51 m. While main channel length is 60.46 km, total streams length within watershed are 346.72 km. The 2002 HUC 10 watershed land use estimates 0.09, 0.17 and 0.05% of watershed as water, wetland and wetland forest, respectively. Deciduous forest, ungrazed grass, grazed grass, CRP grassland, and alfalfa are 2.71, 10.87, 2.52, 1.7, and 1.84%, respectively. Corn and soybeans, and other row crops are 41.16, 32.95, and 0.43%, respectively. Common/industrial, residential, and barren land are 1.67, 1.27, and 0.06%, respectively. Land cover map is shown in Figure 1(bottom). The study area has average annual rainfall of 804 mm (arithmetic mean of 2007, 2008, 2009, and 2010 years). The watershed data were obtained from Natural Resources Geographic Information System (NRGIS) library. The library is maintained by GIS section of the Iowa Department of Natural Resources (IDNR). Data of stream flow were obtained from the U.S. Geological Survey gaging station 05470500 (Lat 42°01'23", long 93°37'49") on Squaw Creek in Ames. The air temperature and precipitation data of Ames (Lat 42° 01'48", long 93° 04'48") were obtained from Iowa Environmental Mesonet (IEM), Agronomy Department, Iowa State University, USA. To quantify the manure applied in the watershed, we used manure application map 2006, which was prepared by the Iowa Department of Natural Resources (DNR) that describes the area which potentially receives the manure. The map uses the CAFO locations, manure production, and manure application rates to calculate the land area receiving animal manure as fertilizers.

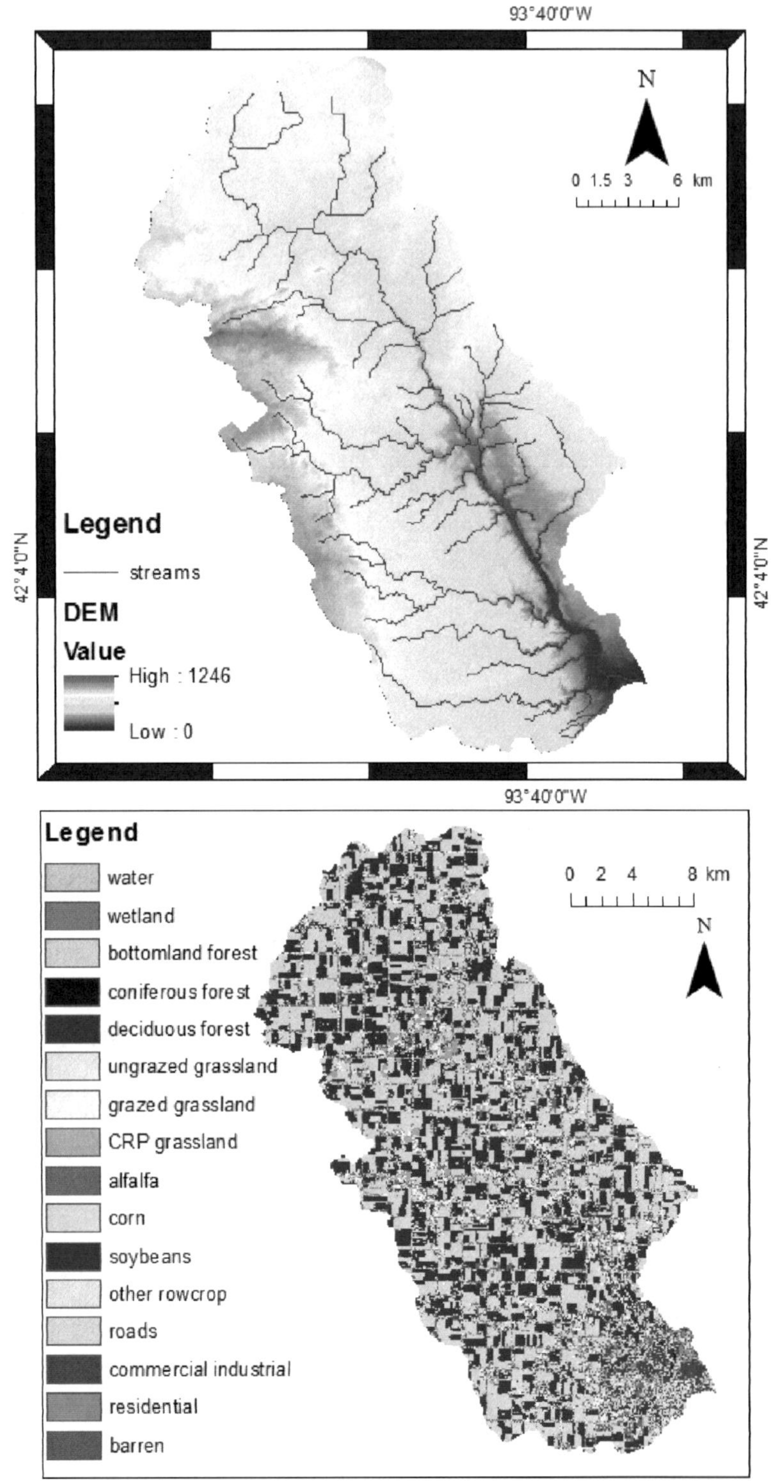

Figure 1. Study area maps: DEM and streams (top) and land cover (bottom)

3. Methods

The SWAT model, which was used in this study, requires input parameters such as land cover area, soil type, rainfall, temperature, slopes, and streams of the watershed. Using the watershed characteristics and hydrology, the model predicts stream flow. The SWAT (a river basin/watershed scale hydrological model), developed and supported by United State Department of Agriculture (USDA) Agricultural Research Service, is available freely at http://swat.tamu.edu/. The model has been extensively used in predicting daily/monthly stream flow and water quality (i.e., nutrients, pesticides, sediment) around the world [11]. Numerous studies have exploited the SWAT model for understanding the impacts of land use management practices on water quantity as well as water quality [11, 12, 13].

The SWAT model divides a watershed into multiple sub-basins. These sub-basins are called hydrological response units (HRU). The HRU includes the homogenous land use and management, soil types, and slopes. The implementation of the SWAT model for predicting stream flow and nutrients is described extensively [12, 13, 14]. Implementation and application of SWAT model for predicting the bacteria concentrations is described elsewhere [8, 9, 15]. The input parameters such as manure application rate, crop land receiving manure, crop rotation, land cover, erosion rate, and particle attached and non- particle attached bacteria are used in the SWAT to estimate bacteria loading into the streams. The SWAT model estimates overland bacteria transport from the crop land to streams. While predicting overland bacteria transport, the model involves simulating bacteria in surface runoff, bacteria attached to sediment in surface runoff, and bacteria lag in surface runoff. The details of the overland bacteria transports processes and related parameters are described elsewhere [8, 9, 10, 14, 15]. In this study we used the modified SWAT model, which is reported elsewhere in details [10] (http://lib.dr.iastate.edu/cgi/viewcontent.cgi?article=3862&context=etd).

4. Results and Discussion

Figure 2 shows measured and predicted stream flow. Results show that the predictions of daily stream flow by the model are well matched with the daily stream flow observations at gaging station. Stream

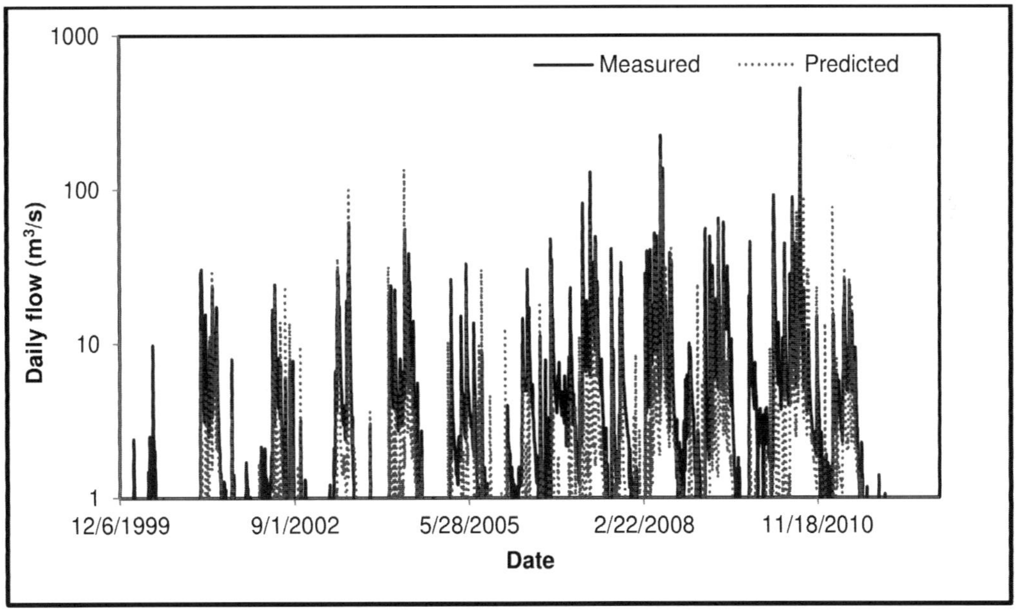

Figure 2. Observed and predicted daily stream flow

flow was simulated from 2000 to 2011, and observed stream flow data (shown in Figure 2) were obtained from USGS gaging stations. The coefficient of determination (r^2) and Nash-Sutcliffe's efficiency coefficient (*NSE*) values were estimated to verify the model predictions. The r^2 and *NSE* values between predicted and measured daily stream flow were r^2 = 0.42 and *NSE* = 0.39. The correlation coefficient (r) between predicted and measured daily flow was 0.65. While r^2 and *NSE* values for daily flows were lower, for average monthly daily flow these values were considerably higher [10]. The model was also used to predict bacteria concentrations (*E. coli* levels) in streams, which is shown in Figure 3. As shown in the figure, increase and decrease in bacteria concentrations followed the stream flow. These results are similar to findings by previous studies, which have shown that bacteria concentrations in the stream increases with elevated stream flow [10, 8, 9]. To verify the model predictions, we compared predicted *E. coli* levels with the measured *E. coli* levels in stream water. Comparison between measured and predicted data indicated that approximately 82% of the predictions were within one order of magnitude of the observed values [10]. These results are satisfactory considering the uncertainties and complexities involved in predicting stream bacteria levels at the watershed scale. While this study summarizes a tool and method describing general guidelines, published studies, application methods, and study and data sources to implement the SWAT model for predicting stream bacteria levels, we recommend readers studying SWAT model manual and relevant papers. Further studies understanding how the manure application rates in the watershed, variation in rainfall, and temperature can potentially impact in-stream bacteria concentrations are required for identifying the efficient manure management practices, which can support in mitigating bacteria concentrations in the streams.

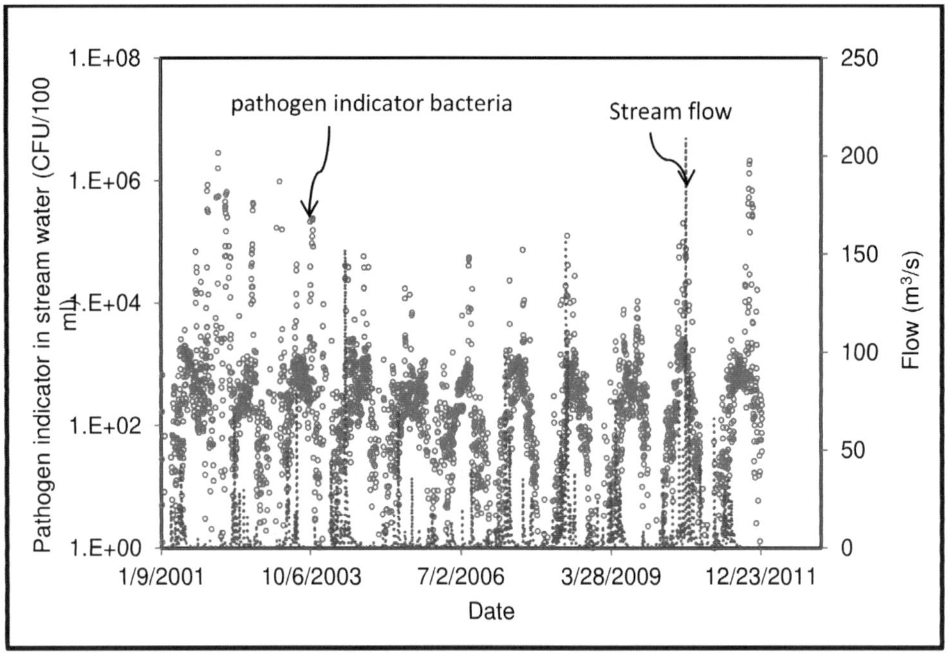

Figure 3. Predicted pathogen indicator bacteria and stream flow.

5. Conclusions

Here we have described briefly the hydrological modelling approach to simulate animal manure bacteria transport from cropping land receiving manure as fertilizers to stream water. The SWAT model, input data, methods, published studies on model application, and simulation processes are summarized; however, to implement the model additional studies will be required, which are recommended in this paper. We suggest readers to review SWAT manual and input and output documentation, which is available freely, prior to exploiting the model. While here we have presented the results of stream flow and bacteria predictions of a single watershed, readers are encouraged to review multiple studies which are already published, describing the SWAT model applications at watershed scale, for predicting bacteria levels as well as nutrient concentrations in streams.

References

[1] US Environmental Protection Agency (US EPA) (2012). Impaired waters and total maximum daily loads. http://water.epa.gov/lawsregs/lawsguidance/cwa/tmdl/index.cfm (accessed on 4.12.11).

[2] Soupir, M.L., Mostaghimi S., Yagow, E.R., Hagedorn, C. and Vaughan, D.H. (2006) Transport of fecal bacteria from poultry litter and cattle manures applied to pastureland. E. coli and Enterococci in dairy cowpats. Water Air Soil Pollut. 169, 125–136.

[3] Armstrong, S.D., Smith, D.R., Joern, B.C., Owens, P.R., Leytem, A.B., Huang, C. and Adeola, L. (2009) Transport and Fate of Phosphorus during and after Manure Spill Simulations. J. Environ. Qual. 39, 345-352.

[4] Centner, T.J., Feitshans, T.A. (2006) Regulating manure application discharges from concentrated animal feeding operations in the United States. Environmental Pollution 141, 571-573.

[5] Jagai, J. S., Griffiths, J.K., Webb, P., Naumova, P. (2010) Patterns of Protozoan Infections: Spatiotemporal Associations with Cattle Density. EcoHealth 7(1):33-46.

[6] Lichtfouse, E., Armstrong, S. D., Smith, D. R., Owens, P. R., Joern, B., Williams, C. (2010) Manure Spills and Remediation Methods to Improve Water Quality. Pages 201-215 in Genetic Engineering, Biofertilisation, Soil Quality and Organic Farming. Vol. 4. Springer Netherlands.

[7] Pandey, P.K., Soupir, M.L., Haddad, M., Rothwell, J.J. (2012) Assessing the impacts of watershed indexes and precipitation on spatial in-stream E. coli concentrations. Ecol. Indic. 23, 641-652.

[8] Coffey, R., Cummins, E., Bhreatyhnach, N., Flaherty, V.O., Cormican, M. (2010) Development of a pathogen transport model for Irish catchments using SWAT. Agr. Water Manage. 97, 101–111.

[9] Kim, J.W., Pachepsky, Y.A., Shelton, D.R. and Coppock, C. (2010) Effect of streambed bacteria release on E. coli concentrations: Monitoring and modeling with the modified SWAT. Ecol. Model. 221(12), 1592-1604.

[10] Pandey, P.K. (2012) Modelling in-stream E. coli cocentraitons. PhD dessertation, Iowa State University, Ames, Iowa, USA.

[11] Gassman, P.W., Reyes, M., Green, C.H., Arnold, J.G. (2007) The Soil and Water Assessment Tool: Historical development, applications, and future directions. Trans. ASABE 50, 1211–1250.

[12] Jha, M., Schilling, K. E., Gassman, P. W., Wolter, C. F. (2010) Targeting land use change for nitrate-nitrogen load reductions in an agricultural watershed. J. Soil Water Conserv. 65(6), 342-352.

[13] Jha, M.K., Wolter, C.F., Schilling, K.E., Gassman, P.W. (2010) Assessment of total maximum daily load implementation strategies for nitrate impairment of the Raccoon River, Iowa. J. Environ. Qual. 39(4), 1317–1327.

[14] Neitsch, S.L., Arnold, J.G., Kiniry, T.R. and Williams, J.R. (2005) Soil and Water Assessment Tool: Theoretical documentation. Texas Water Resources Inst., College Station.

[15] Parajuli, P.B., Mankin, K.R., Barnes, P.L. (2009) Source specific fecal bacteria modeling using soil and water assessment tool model. Bioresour. Technol. 100, 953–963.

IOP Conf. Series: Earth and Environmental Science **19** (2014) 012015 doi:10.1088/1755-1315/19/1/012015

An input-output approach to analyze the ways to increase total output of energy sectors: The case of Japan

Ubaidillah Zuhdi

Department of Industrial Administration, Tokyo University of Science, Noda 278-8510, Japan
School of Business and Management, Institut Teknologi Bandung, Bandung 40132, Indonesia

ubaidillah.zuhdi@gmail.com

Abstract. The purpose of this study is to analyze the ways to increase total output of Japanese energy sectors in future time. In this study, Input-Output (IO) analysis is employed as a tool of analysis. This study focuses on petroleum refinery products and non-ferrous metals as analyzed sectors. The results show that positive impact observed in export and outside households consumption modifications while opposite impact is given by modification of import. The recommendations suggested based on these results are Japanese government should make breakthroughs so analyzed sector's export activities can increase and they have to careful in conducting import activities related to these sectors.

1. Introduction

Humans need energy in order to conduct their activities. This dependency will continue not only in micro but also macro aspects. For example, in micro scale, people need energy in conducting daily tasks such as work and sport. In macro scale, countries direct and indirectly need energy, represented by their energy sectors, in order to execute their economic activities such as making new workplaces and enhancing the education level of society. These arguments indicate that energy is also economically important.

There are many previous studies investigated the relationship between energy and economic aspects. [1] analyzed the relationship between changes of oil price and real Gross Domestic Product (GDP) which the focus was US economy. [2] explained the connections between Euro area's inflation and oil prices. Besides, [3] explored the effect of oil price shocks on growth of economy of some Middle East and North Africa (MENA) countries. The study focuses on the ways to increase the amount of energy, represented by total output of energy sectors, using economic analysis, however, still lack. The kind of study, actually, is needed in order to know, from point of view of economy, the improvements should be done in enhancing the sectors. This study is conducted in order to fulfill the gap.

The purpose of this study is to analyze the ways to increase energy sector's total output in future time. In this study, the object of study is Japan, one of developed country in the world. This study employs Input-Output (IO) analysis as a tool of analysis. More specifically, this study employs demand-pull IO quantity model in achieving the purpose.

2. Methodology

As mentioned in introduction, this study uses demand-pull IO quantity model in order to know the ways to enhance total output of Japanese energy sectors in future time. Following explanation is provided to describe the methodology of this study. The first step is to describe Japanese energy sectors used in this study. Table 1 shows these sectors. The focused energy sectors are chosen based on results of previous study. [4] explained that petroleum refinery products and non-ferrous metals sectors had positive gross output changes during 1995-2005. The positive changes in gross output of one sector indicate that this sector has positive growth in specific period. In other words, this sector gives strong impact in national economy in the period. Further, higher increase on total output of this sector will generate higher enhancement on national economy in future time. This study then follows their results, namely exposing petroleum refinery products and non-ferrous metals as analyzed sectors.

Table 1. Japanese energy sectors used in this study.

No.	Sector number	Sector name
1	6	Metallic ores
2	7	Non-metallic ores
3	8	Coal mining, crude petroleum and natural gas
4	26	Final chemical products, n.e.c.
5	27	Petroleum refinery products
6	28	Coal products
7	36	Pig iron and crude steel
8	37	Steel products
9	38	Steel castings and forgings and other steel products
10	39	Non-ferrous metals
11	40	Non-ferrous metal products
12	41	Metal products for construction and architecture
13	42	Other metal products

(Source: [4] with slight modifications)

The second step is to do the calculation in order to know the ways to increase total output of analyzed sectors. Demand-pull IO quantity model is used in this calculation. According to [5], following equation describes this model:

$$\mathbf{x}^1 = \mathbf{L}^0\mathbf{f}^1 \tag{1}$$

where x, L, and f are matrices of total output of industrial sectors, Leontief inverse, and final demand of industrial sectors, respectively. *0* and *1* indicate present and future periods, respectively. Above equation explains that final demand modifications will give impact to total output of sectors. In other words, the ways that are conducted in order to give impact on total output of one sector in future time can be represented by the final demand changes. Table 2 describes the scenarios of final demand modification used in this study. Modified 2005 IO table of Japan is used as a base of calculation.

"Whole sectors change" and "pure change" conditions, actually, are considered in this calculation. Former word describes the condition where modifications of final demand are addressed to all Japanese industrial sectors while latter one only focuses on analyzed sectors. In this study, former one will be called condition A while condition B is used in describing latter condition. Recommendations then suggested based on calculation results.

Table 2. The scenarios of final demand modification used in this study.

Component of final demand	Scenario 1	Scenario 2	Scenario 3
	Export modification	Import modification	Outside households consumption modification
Export	Rises 30%	Constant	Constant
Import	Constant	Rises 30%	Constant
Outside households consumption	Constant	Constant	Rises 30%

(Source: [6] with slight modifications)

3. Results and discussion

Figures 1 and 2 describe total output of petroleum refinery products and non-ferrous metals sectors for each scenario in condition A, respectively. On the other hand, figures 3 and 4 expose total output of these sectors for each scenario in condition B. In this study, *t* and *t+1* indicate present and future times, respectively.

Based on information in the figures, for both conditions, similar patterns appear in analyzed sectors. These patterns can be described as follows. First, positive impact, represented by increasing on total output in future time, observed in scenarios 1 and 3. Second, negative impact, represented by decreasing on total output in future time, is given by import modification.

The figures also show that scenario 1 gives higher positive impact than scenario 3 in analyzed sectors. Therefore, this scenario should be prioritized by Japanese government in order to increase total output of these sectors. In other words, Japanese government should make breakthroughs so these sector's export activities can increase. For example, improving the quality of refinery products of petroleum and making the competitive prices for these products.

Another aspect should be considered by Japanese government is import activities related to analyzed sectors. These activities, actually, give negative impact to these sectors. Therefore, they have to careful in conducting these activities. Import restriction related to products of the sectors is a good example.

4. Conclusions and further researches

The purpose of this study is to analyze the ways to increase total output of Japanese energy sectors in future time. In this study, demand-pull IO quantity model is employed as a tool of analysis. This study focuses on petroleum refinery products and non-ferrous metals as analyzed sectors. Three scenarios of final demand modification are used in the calculation of this study. These are (1) export, (2) import, and (3) Outside households consumption modifications. This study uses modified 2005 IO table of Japan as a base of calculation. "Whole sectors change" and "pure change" conditions are also considered in the calculation.

The results show that positive impact, represented by increasing on total output in future time, observed in scenarios 1 and 3 while opposite impact is given by modification of import. Besides, the results expose that scenario 1 gives higher positive impact than scenario 3 in analyzed sectors. This phenomenon appears in both conditions. The recommendations suggested based on these results are Japanese government should make breakthroughs so analyzed sector's export activities can increase and they have to careful in conducting import activities related to these sectors. More specifically, they should (1) improve the quality of refinery products of petroleum, (2) make the competitive prices for these products, and (3) do the restriction for import activities related to the products.

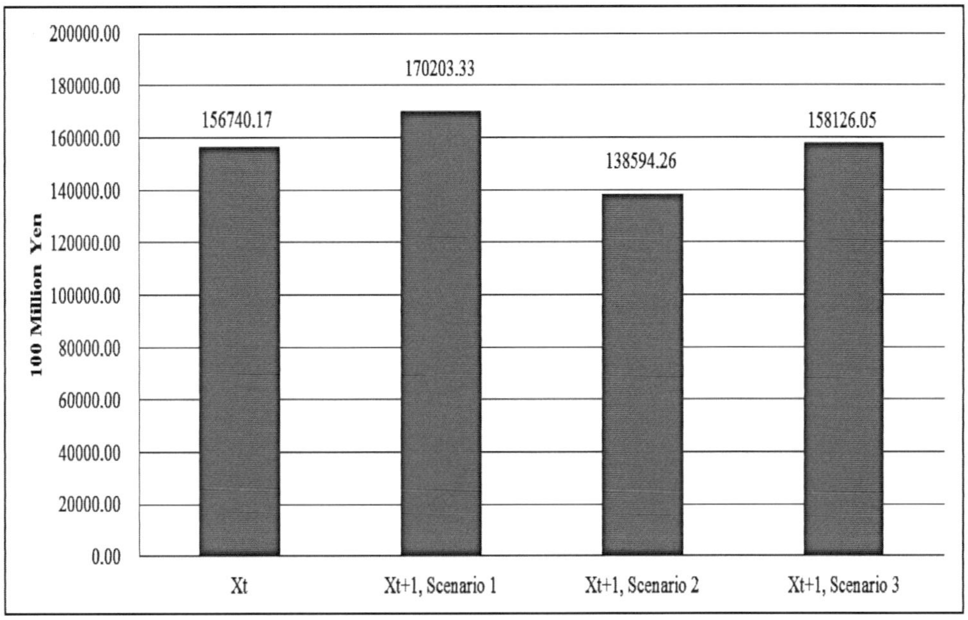

Figure 1. Total output of petroleum refinery products sector for each scenario in condition A.

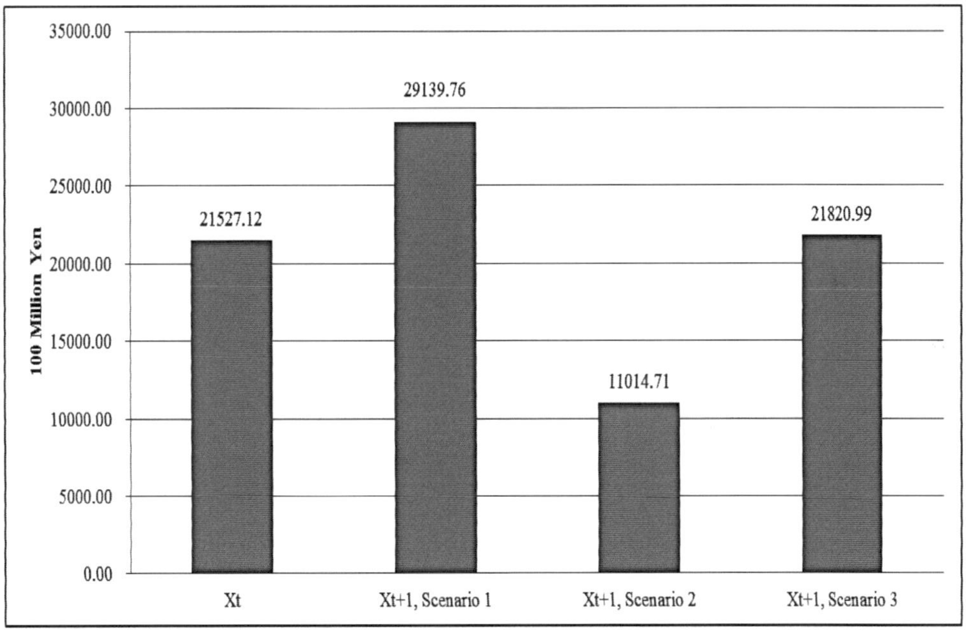

Figure 2. Total output of non-ferrous metals sector for each scenario in condition A.

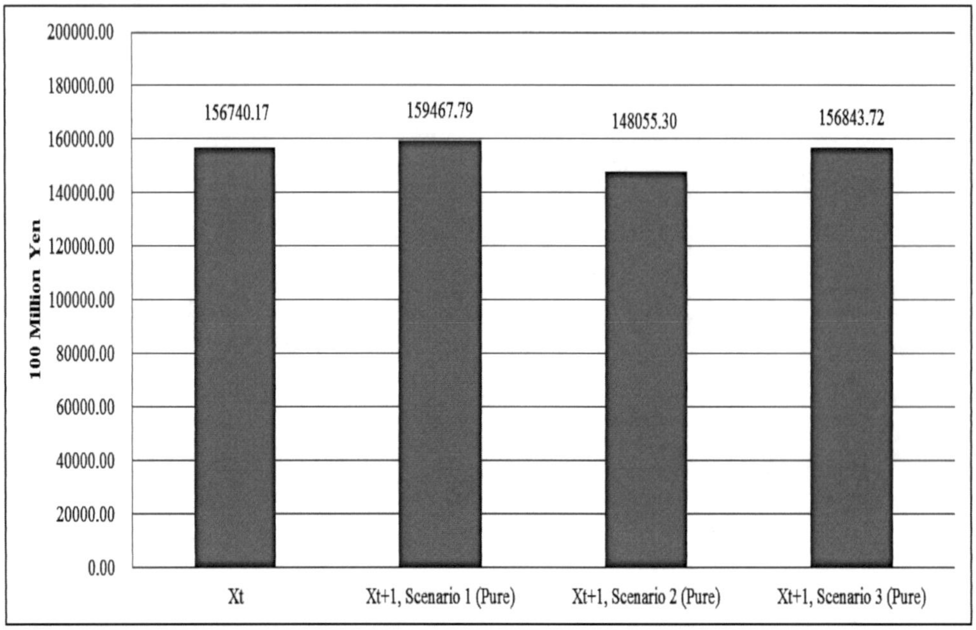

Figure 3. Total output of petroleum refinery products sector for each scenario in condition B.

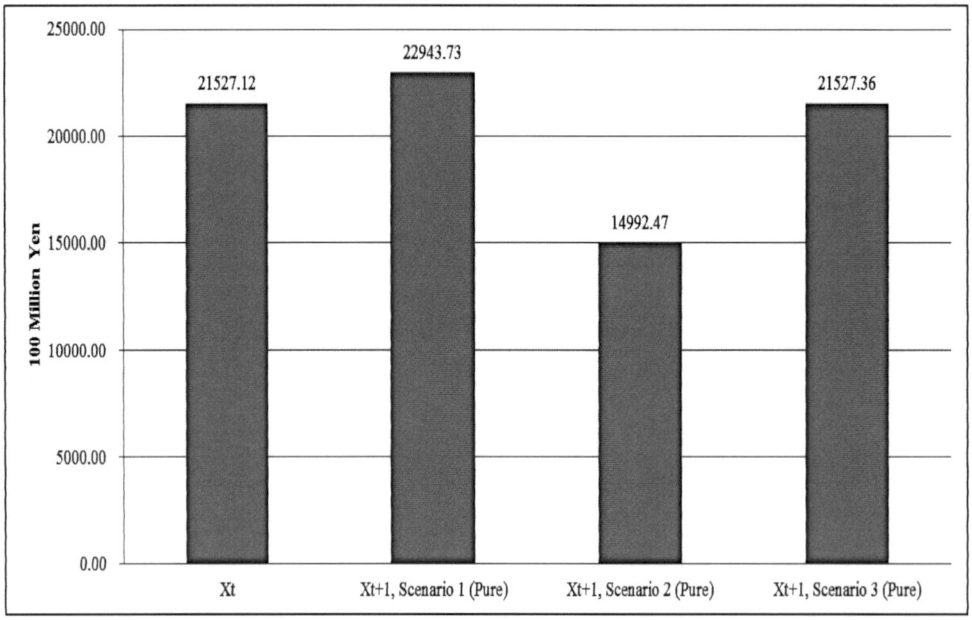

Figure 4. Total output of non-ferrous metals sector for each scenario in condition B.

This study, however, only focuses on petroleum refinery products and non-ferrous metals sectors. Other Japanese energy sectors analysis might be generate different results compared with this study. Therefore, this analysis is a suggested further research from this study. Besides, this study also suggests international comparison related to this topic. This comparison will describe the characteristics of energy sectors in compared countries including the difference and / or similarity on the ways to increase total output of these sectors.

References

[1] Jimenez-Rodriguez R 2009 Oil price shocks and real GDP growth: testing for non-linearity *The Energy Journal* **30** 0124
[2] Jacquinot P, Kuismanen M, Mestre R and Spitzer M 2009 An assessment of the inflationary impact of oil shocks in the euro area *The Energy Journal* **30** 4984
[3] Berument H, Ceylan N B and Dogan N 2010 The impact of oil price shocks on the economic growth of selected MENA countries *The Energy Journal* **31** 149176
[4] Zuhdi U and Mori S 2012 Analyzing the gross output change of energy sector industry in Indonesia between 1990 and 2005 by decomposition analysis *Proc. of The 3rd International Association for Energy Economics Asian Conference (Kyoto, Japan, 20–22 February 2012)*
[5] Miller R E and Blair P D 2009 *Input-Output Analysis: Foundations and Extensions* (Cambridge: University Press)
[6] Zuhdi U, Prasetyo A D and Sianipar C P M 2013 Analyzing the dynamics of total output of Japanese creative industry sectors: An input – output approach *Procedia Economics and Finance* **5** 827835

IOP Conf. Series: Earth and Environmental Science **19** (2014) 012016 doi:10.1088/1755-1315/19/1/012016

Analyzing the impacts of final demand changes on total output using input-output approach: The case of Japanese ICT sectors

Ubaidillah Zuhdi

Department of Industrial Administration, Tokyo University of Science, Noda 278-8510, Japan
School of Business and Management, Institut Teknologi Bandung, Bandung 40132, Indonesia

ubaidillah.zuhdi@gmail.com

Abstract. The purpose of this study is to analyze the impacts of final demand changes on total output of Japanese Information and Communication Technologies (ICT) sectors in future time. This study employs one of analysis tool in Input-Output (IO) analysis, demand-pull IO quantity model, in achieving the purpose. There are three final demand changes used in this study, namely (1) export, (2) import, and (3) outside households consumption changes. This study focuses on "pure change" condition, the condition that final demand changes only appear in analyzed sectors. The results show that export and outside households consumption modifications give positive impact while opposite impact could be seen in import change.

1. Introduction

The utilization of Information and Communication Technologies (ICT) can be easily seen in recent times. For example, people use mobile phone when doing communication with others. Another example is the usage of computer and internet by people in their daily activities. This phenomenon, actually, can be observed almost everywhere and every day. In other words, now ICT have been important aspect in human life.

There are many previous studies discussed about ICT. [1] investigated the impact of Foreign Direct Investment (FDI) and ICT on productivity growth. Their analysis was based on panel data of 42 developed and developing countries on the period between 1993 and 2001. [2] analyzed the role of ICT sectors in national economic structural changes of Indonesia during 1990-2005. They employed structural decomposition analysis as a tool of analysis. [3] conducted the international comparison in order to know the role of ICT in structural changes of national economy of analyzed countries, Indonesia and Japan. Similar with previous study, they also used structural decomposition analysis in doing analysis. The study focuses on impacts of changes of exogenous element on total output of ICT sectors in one country, however, is still lack. The kind of study, actually, can open the opportunity to increase total output of these sectors in future time. This study is conducted in order to fulfill the gap.

The purpose of this study is to analyze the impacts of changes of exogenous element, final demand, on total output of ICT sectors in future time. This study focuses on Japan, one of developed country in the world. The Input-Output (IO) analysis is employed as a tool of analysis in this study.

2. Methodology

As described in introduction, the purpose of this study is to analyze the impacts of final demand changes on total output of Japanese ICT sectors in future time. IO analysis is employed as an analysis tool. More specifically, this study uses demand-pull IO quantity model in order to achieve the purpose. Following explanation describes about methodology of this study.

The first step is to describe Japanese ICT sectors used in this study. Table 1 shows these sectors. The second step is to conduct the calculation in order to know the impacts of modifications of final demand to total output of the sectors. Demand-pull IO quantity model is used in this calculation. According to [4], following equation describes this model:

$$\mathbf{x}^1 = \mathbf{L}^0\mathbf{f}^1 \quad (1)$$

where x, L, and f are matrices of total output of industrial sectors, Leontief inverse, and final demand of industrial sectors, respectively. *0* and *1* indicate present and future times, respectively. Table 2 describes the scenarios of final demand modification used in this study. Modified 2005 IO table of Japan is used as a base of calculation. This study focuses on "pure change" condition, the condition that final demand changes only appear in analyzed sectors. The recommendations are then suggested based on calculation results.

Table 1. Japanese ICT sectors used in this study.

No.	Sector number	Sector name
1	72	Communication
2	73	Broadcasting and information services
3	80	Advertising, survey, and information services

(Source: [3])

Table 2. The scenarios of final demand modification used in this study.

Component of final demand	Scenario		
	1	2	3
	Export modification	Import modification	Outside households consumption modification
Export	Rises 30%	Constant	Constant
Import	Constant	Rises 30%	Constant
Outside households consumption	Constant	Constant	Rises 30%

(Source: [5] with slight modifications)

3. Results and discussion

Figures 1, 2, and 3 describe total output of Japanese ICT sectors for each scenario. All analyzed sectors, based on information in these figures, have similar patterns, namely scenario 1 and 3 gives positive impact while opposite impact observed in the calculation using scenario 2. A subtle difference, however, appears on former one. In communication sector, scenario 3 gives higher positive impact than scenario 1 while opposite situation happened in broadcasting and information services and advertising, survey, and information services sectors.

Above phenomenon indicates that import activities related to ICT products would decrease total output of Japanese ICT sectors in future time. In other words, these activities should be done carefully and mitigated. Japanese government should active in endorsing this restriction. Making regulation supports the mitigation is a good example.

Besides mitigating the import activities related to ICT products, to consider the ICT export and domestic markets are also important things. In other words, the actions to excite these markets should also be done in order to increase total output of Japanese ICT sectors. The examples of these actions are making new types and improving the quality of ICT products. Making the prices of these products competitive is also another example. Certainly, these actions can be combined in order to get optimal results.

4. Conclusions and further researches

The aim of this study is to analyze the impacts of final demand changes on total output of Japanese ICT sectors in future time. This study employs demand-pull IO quantity model in achieving the purpose. There are three final demand changes used in this study, namely (1) export, (2) import, and (3) outside households consumption changes. The "pure change" condition, condition that final demand changes only appear in analyzed sectors, is a main circumstance in this study.

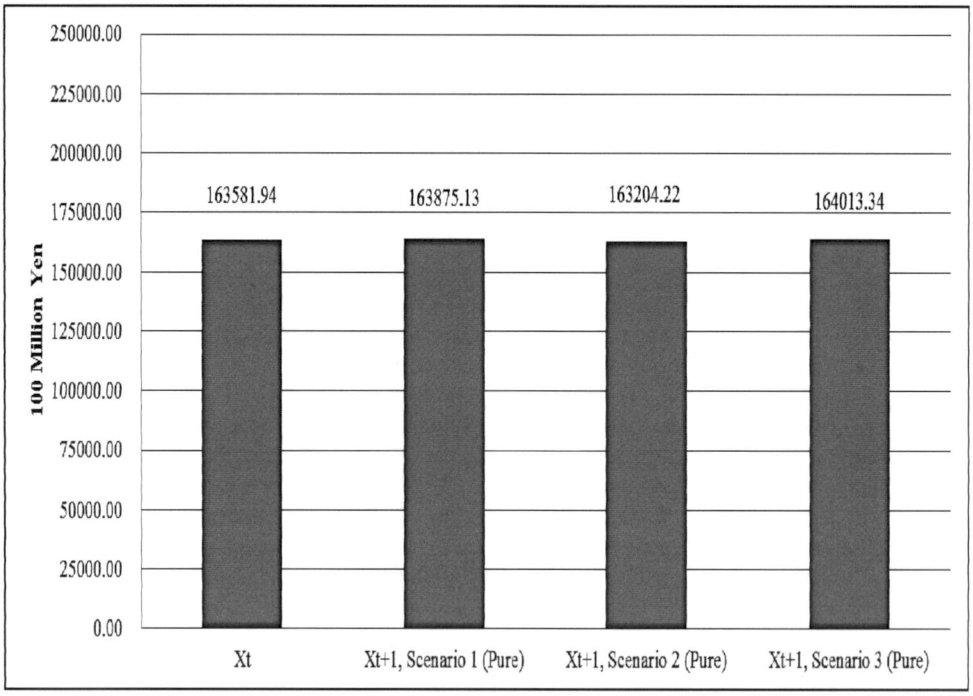

Figure 1. Total output of communication sector for each scenario.

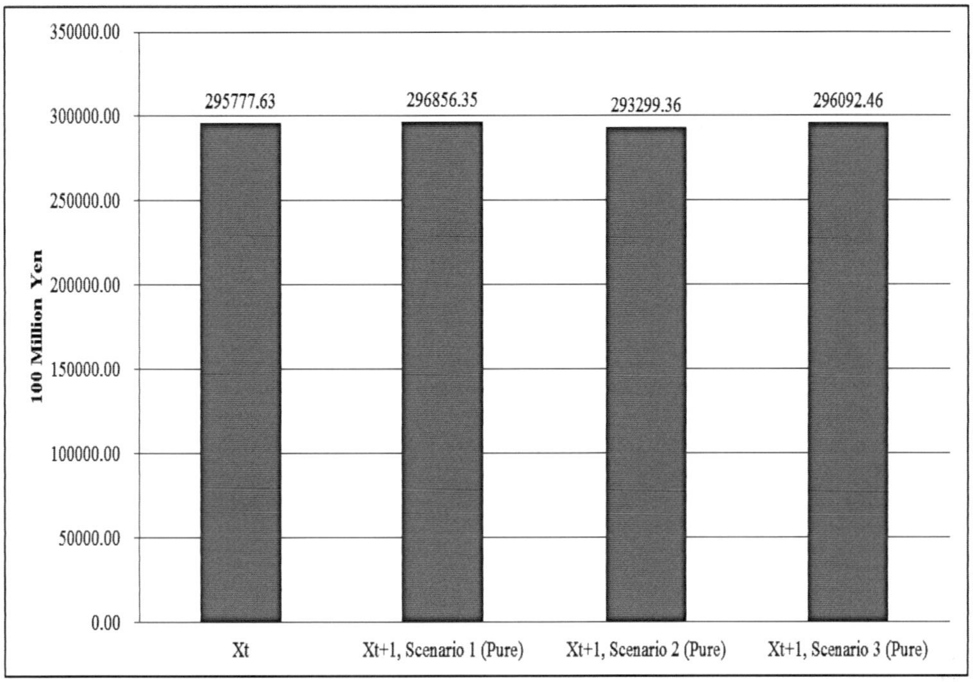

Figure 2. Total output of broadcasting and information services sector for each scenario.

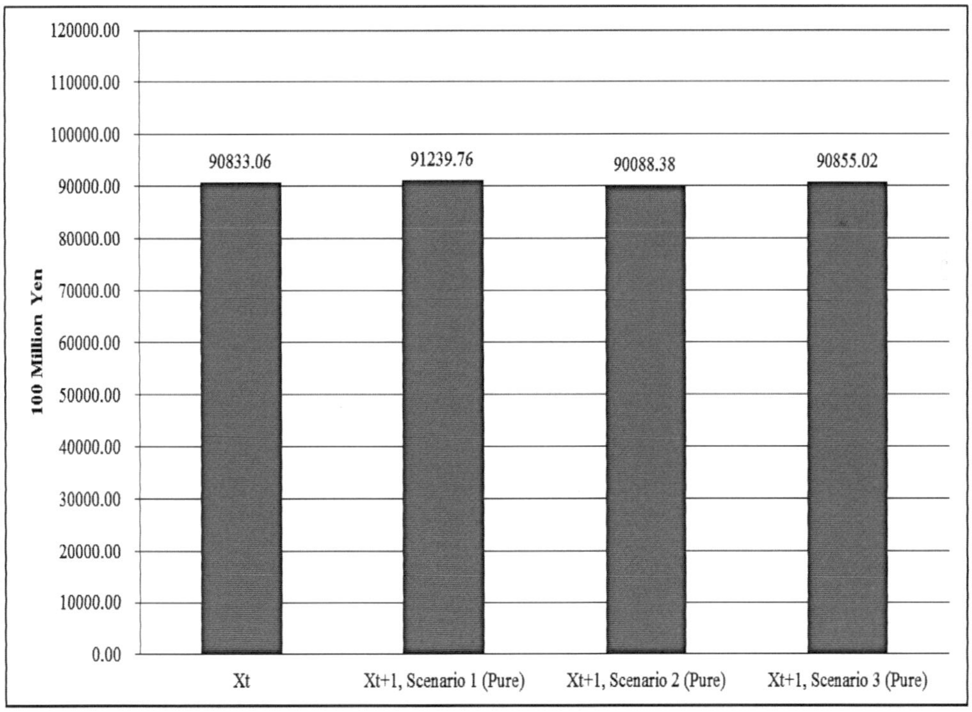

Figure 3. Total output of advertising, survey, and information services sector for each scenario.

The results show that export and outside households consumption modifications give positive impact while opposite impact could be seen in import change. This phenomenon indicates that import activities related to ICT products should be done carefully and mitigated. Japanese government should active in endorsing this restriction. Making regulation supports this mitigation is a good example. Besides, the actions to excite ICT export and domestic markets should also be done in order to increase total output of Japanese ICT sectors. For example, making new types and improving the quality of ICT products. Making the prices of these products competitive is also another example. The combination of these actions can be done in order to achieve optimal results.

This study, however, only analyzes the impact of final demand changes on total output of Japanese ICT sectors. In other words, deeper analysis on other points, such as analysis of current condition of ICT of Japan and their ICT policies, is not discussed in this study. This analysis, actually, is also important and should be considered in achieving the enhancement of total output of Japanese ICT sectors. Therefore, this kind of analysis is a suggested further research from this study.

Conducting IO multipliers analysis in this topic is also suggested further research from this study. This analysis, actually, will clearly describe the role of industrial sectors in national economy of one country. This study especially recommends to conducting income/employment multipliers analysis in further research because it can show the impact of final demand changes on additional earnings received by employees of specific sector. In other words, this study will show the sectors generate high additional earnings for their employees. Obviously, this analysis can be focused on specific sector, says ICT sector, so historical changes of additional earnings of this sector can be studied. Another suggested further research from this study is an international comparison, especially the comparison between developed and developing countries, related to this topic. This kind of study will show the characteristics of ICT sectors of analyzed countries, especially when final demand changes happened.

References

[1] Dimelis S P and Papaioannou S K 2009 FDI and ICT effects on productivity growth: A comparative analysis of developing and developed countries *European Journal of Development Research* **22** 7996

[2] Zuhdi U, Utomo D S and Alamanda D T 2011 Analyzing the role of ICT sector to the national economic structural change: The case of Indonesia *Jurnal Manajemen Teknologi* **10** 299307

[3] Zuhdi U, Mori S and Kamegai K 2012 Analyzing the role of ICT sector to the national economic structural changes by decomposition analysis: The case of Indonesia and Japan *Procedia-Social and Behavioral Sciences* **65** 749754

[4] Miller R E and Blair P D 2009 *Input-Output Analysis: Foundations and Extensions* (Cambridge: University Press)

[5] Zuhdi U, Prasetyo A D and Sianipar C P M 2013 Analyzing the dynamics of total output of Japanese creative industry sectors: An input – output approach *Procedia Economics and Finance* **5** 827835

Temporal Forest Change Detection and Forest Health Assessment using Remote Sensing

Norsuzila Ya'acob[1,2,*], Aziean Binti Mohd Azize[1], Nur Anis Mahmon[1], Azita Laily Yusof[1,2], Nor Farhana Azmi[1] and Norfazira Mustafa[1]

[1] Faculty of Electrical Engineering, Universiti Teknologi MARA, 40450 Shah Alam Selangor, Malaysia
[2] Wireless Communication Technology (WiCoT), Faculty of Electrical Engineering, Universiti Teknologi MARA, 40450 Shah Alam, Selangor, Malaysia
*Institute of Space Science, Universiti Kebangsaan Malaysia, 43600 Bangi, Selangor, Malaysia.

norsuzila@salam.uitm.edu.my,nuranismahmon@yahoo.com, azita968@salam.uitm.edu.my,farhanaazmi_89@yahoo.com, faziramustafa@yahoo.com

Abstract This paper presents the detection of Angsi and Berembun Reserve Forest change for years 1996 and 2013. Forest is an important part of our ecosystem. The main function is to absorb carbon oxide and produce oxygen in their cycle of photosynthesis to maintain a balance and healthy atmosphere. However, forest changes as time changes. Some changes are necessary as to give way for economic growth. Nevertheless, it is important to monitor forest change so that deforestation and development can be planned and the balance of ecosystem is still preserved. It is important because there are number of unfavorable effects of deforestation that include environmental and economic such as erosion of soil, loss of biodiversity and climate change. The forest change detection can be studied with reference of several satellite images using remote sensing application. Forest change detection is best done with remote sensing due to large and remote study area. The objective of this project is to detect forest change over time and to compare forest health indicated by Normalized Difference Vegetation Index (NDVI) using remote sensing and image processing. The forest under study shows depletion of forest area by 12% and 100% increment of deforestation activities. The NDVI value which is associated with the forest health also shows 13% of reduction.

1. Introduction

Reserve forest in Negeri Sembilan as of 2011 is 154,577 hectare which includes 24% of the state's land cover. There are 21 reserve forests here and among them are Angsi and Berembun with approximately 34,000 hectares in combine. Remote sensing is the science of acquiring information about the earth's surface without actually being in contact with it. Remote sensing technology involve the use of sensor placed on a platform moving at a distance from earth's surface and it can be used for collecting data of the earth for the purpose of inventorying and monitoring. Remote sensing is done by receiving and recording energy that is either emitted or reflected by the surface of the earth, so there must be a source of electromagnetic energy, a target, and a sensor.

Reflectance is the percentage of light hitting a material that is then reflected by that material. These patterns of reflectance and absorption across wavelengths can uniquely identify certain materials. A chemical compound in leaves called chlorophyll strongly absorbs radiation in the red and blue wavelengths but reflects green wavelengths. Leaves appear green when chlorophyll content is at its maximum. When there is less chlorophyll in the leaves, there is less absorption and proportionately more reflection of the red wavelengths, making the leaves appear yellow. The internal structure of healthy leaves act as excellent diffuse reflectors of near-infrared wavelengths. The sensor receives the reflected energy and it is transmitted to a receiving and processing station where it is converted into digital data which are then converted into images using computer models. In Malaysia command operational satellite used are Radarsat, Landsat, SPOT, Quick Bird and IKONOS. Finally, the images are interpreted to extract information about the target.

Normalized Difference Vegetation Index (NDVI) shows patterns of vegetation growth by indicating the quantity of actively photosynthesizing biomass on a landscape [1]. Measuring and monitoring NDVI is one way that scientists can determine how healthy or unhealthy vegetation may be. There are many factors that can cause the disruption of the health of forests. These can be categorized into living (biotic) and non-living (abiotic) factors. Biotic agents include fungi, bacteria, viruses, insects, and larger animals. Non-living factors are related to weather, water, soil conditions, mechanical agents, and chemicals such as atmospheric pollution [2]. The main goal of this project is to detect forest temporal change using remote sensing application. The remote sensing images are processed and analyzed using Erdas Imagine and ArcGIS software. The NDVI process enables the health of the forest for both years to be analyzed and compared to see their changes over 17 years.

2. Methodology

The method flow of performing this project is divided into several parts. They are:

2.1 Data Collections

Data and information are collected from various sources, namely Malaysian Remote Sensing Agency (Macres), Department of Environment and Forestry Department of Negeri Sembilan. Satellite images used in this study are Landsat-5 TM image year 1996 and SPOT 5 image year 2013 with 2 points latitude and longitude of 2.7, 101.9 and 2.6, 102.2. The images are given with compliment from Macres for this project.

2.2 Image Processing

The raw satellite images were first being pre-processed with radiometric and geometric corrections using Erdas Imagine. Radiometric correction is a process to replace missing or bad lines and to reduce haze. Geometric correction or rectification is the process of projecting the data onto a plane and making it conform to a map projection system. Georeferencing means assigning map coordinates to the image data. Then the images were classified to give a clear view of forest area and forest depletion area in term of deforestation. Supervised classification is identification in the imagery homogeneous representative samples of the different surface cover types or information classes of interest. The 1996 and 2013 images were supervised the categorization of the set of specific classes that are forest, water, deforestation and urban. The images were subset to focus on the study area only. Both images that have been preprocessed and classified were then compared and the resultant image was to be study and analyzed to see the temporal changes between them. Image analyzation isdone using ArcGIS software.

2.3 NDVI Process

This process is done to get the value of NDVI of the forest for year 1996 and 2013 and to compare the NDVI value of each image. NDVI values show how the quality and relative health of the forest.

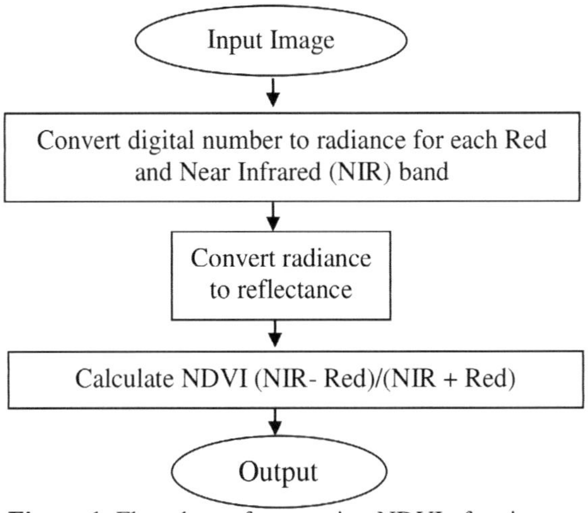

Figure 1. Flowchart of processing NDVI of an image

To investigate the health of the forest, NDVI process was done for both images. The NDVI is the difference of near-infrared (NIR) and visible red reflectance values normalized over reflectance. Before getting the NDVI image, the raw image which is in digital number must be changed to radiance image. The following general equation is used to convert Digital Numbers (DNs) to radiance (L):

$$L_{\lambda} = \text{"gain"} * QCAL + \text{"bias"} \tag{1}$$

This is also expressed a

$$L_{\lambda} = \left(\left(LMAX_{\lambda} - LMIN_{\lambda} \right) / \left(QCALMAX - QCALMIN \right) \right) * \left(QCAL - QCALMIN \right) + LMIN_{\lambda} \tag{2}$$

Where
L_{λ} = Spectral Radiance at the sensor's aperture
$LMIN_{\lambda}$ = the spectral radiance that is scaled to QCALMIN
$LMAX_{\lambda}$ = the spectral radiance that is scaled to QCALMAX
QCALMIN = the minimum quantized calibrated pixel value (corresponding to $LMIN_{\lambda}$) in DN
QCALMAX = the maximum quantized calibrated pixel value (corresponding to $LMAX_{\lambda}$) in DN
QCAL= the quantized calibrated pixel value in DN

The process of changing DN image to radiance image is built in Erdas Imagine model maker. The process is repeated to get radiance image for both year 1996 and 2013. After that, the image in radiance is converted to reflectance while performing atmospheric correction as well. The calculation is as follows:

$$\rho_p = \frac{\pi \,.L\lambda \,.d^2}{ESUN\lambda \;.\cos\theta s} \tag{3}$$

Where
ρ_p = Unitless planetary reflectance
L_{λ} = Spectral radiance at the sensor's aperture
d = Earth-Sun distance in astronomical units
$ESUN_{\lambda}$ = Mean solar exoatmospheric irradiances
$\cos\theta_s$= Solar zenith angle in degree

$$\cos\theta_s = 90^{\circ} - \text{sun elevation} - \text{incident angle} \tag{4}$$

The process of changing radiance image to reflectance image is built in Erdas Imagine model maker using formula (3). The process is repeated to get reflectance images for IR and NIR band for both 1996 and 2013 images. After converting both images to reflectance, NDVI process can be proceed in the model maker. Specifically,

$$NDVI = (NIR - RED) / (NIR + RED) \tag{5}$$

To interpret NDVI value, Table 1 shows reference NDVI values for different land areas [3]. NDVI for vegetation range value is from 0.1 to 1. NDVI value of 0.4 to 1 is commonly associated with rainforest area.

Table 1. Reference NDVI values for different land areas

Different Land Areas	NDVI Values	NDVI Values (X)
Water	Negative values	X < 0
Rocks/Sand/Snow	Values Close to Zero	-0.1 < X < 0.1
Greenery/Vegetation	Low Positive Values	0.1 < X < 0.4
Tropical Rain Forest	Values Approaching to 1	X → 1

3. Result

3.1 Image Classification and Change Detection

The study area is classed into several major classes using supervised classification method. The supervised classification is done for both 1996 and 2013 image. The distribution of classes for each image is graphed. These two images are then compared to see the percentage of changes for each class.

The supervised classification of 1996 Landsat-TM 5 image is depicted and the percentage of distribution areas is shown in Figure 2. The classification distribution chart shows the forest area covers 87% of the whole study area while deforestation, urban and water body cover 7%, 4% and 2% of the whole area respectively.

Applying the same classification method, Figure 3 depict classification distribution chart for year 2013. From the distribution chart of Figure 3, forest area covers 77% of whole study area, deforestation covers 14% while urban and water body cover 7% and 2% of the whole area respectively. The depletion of forest in this study area from year 1996 to 2013 is noticeable. The temporal forest change can be seen with depleting of 12% of forest area which is approximately 4080 hectares of forest reduction. On the other hand, deforestation area increased alarmingly up to 100% from 7% to 14%. The urban area and water body are also increased to 75% and 40% respectively.

The differences between percentages of classification distribution for year 1996 and 2013 show the changes of study area.

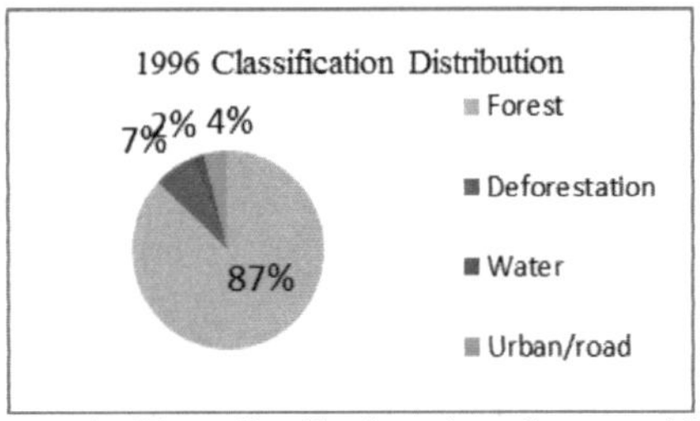

Figure 2. Class distribution chart for year 1996

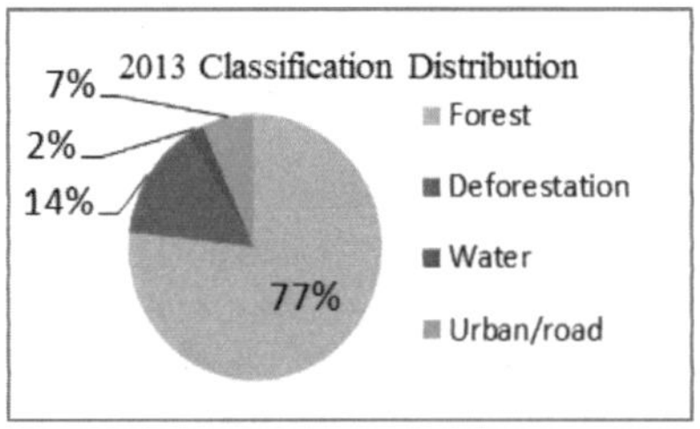

Figure 3. Class distribution chart for year 201

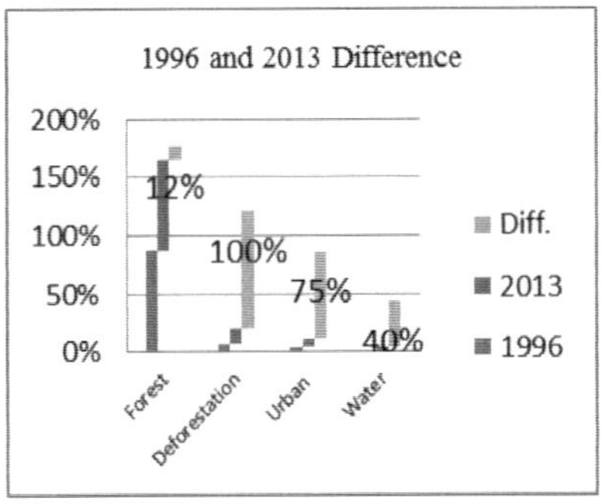

Figure 4. The percentage of class change difference for both images

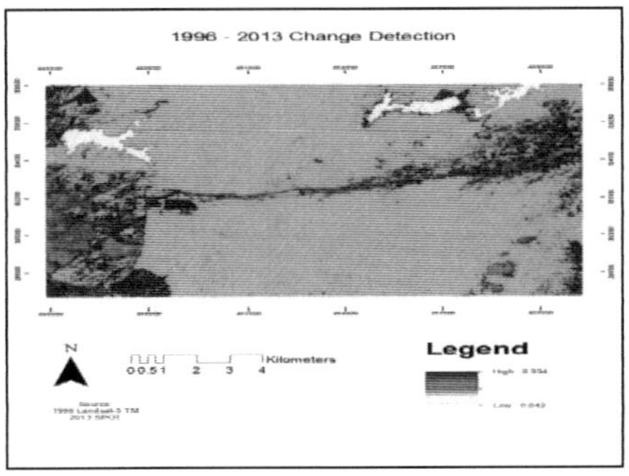

Figure 5. Change detection from 1996 to 201

The change detection for both image are processed and the resultant change detection image is depicted in Figure 5. Figure 4 shows ranges of change based on pixel value from 0 to 1. Light yellow shows little changes, red shows medium changes and blue shows highest changes. The intact forest shows orange in colour which means there is a light change of forest while the deforestation shows blue in colour which means big changes. The urban area shows red means medium changes during the 17 year period.

3.2 NDVI Change Detection and Forest Health

Besides considerable changes of forest area, the health of the forest also shows decrement in term of NDVI. Figure 6 shows the NDVI change image for year 1996 and 2013.

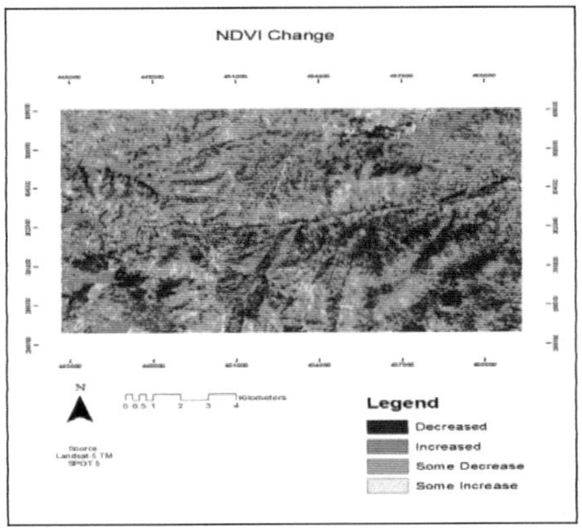

Figure 6. NDVI change for year 1996 and 2013

Figure 6 shows the level of NDVI change detection image. The decrement of NDVI of the forest can be seen with vast brown and light brown colour of the forest area. The increment of NDVI for water body, settlement and urban area which are showed in green colour can be ignored as NDVI is only applicable for vegetation monitoring. It can be seen that almost the whole forest experience NDVI decrement for the 17 years period. The comparison of NDVI statistic for year 1996 and 2013 are shown in Table 2. This comparison shows NDVI statistical value change for both years.

Table 2. NDVI Comparison

Title	NDVI 1996	NDVI 2013	Diff.
Histogram	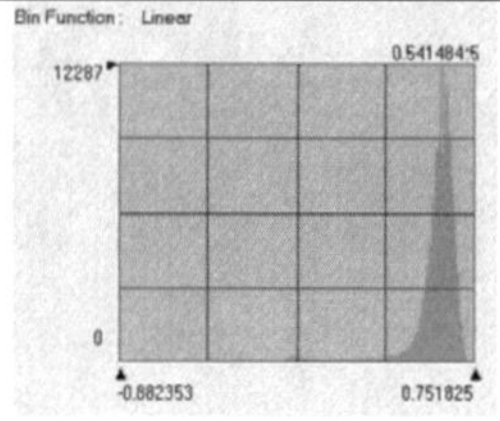	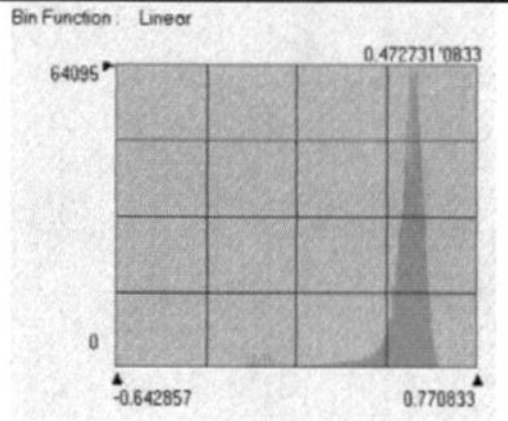	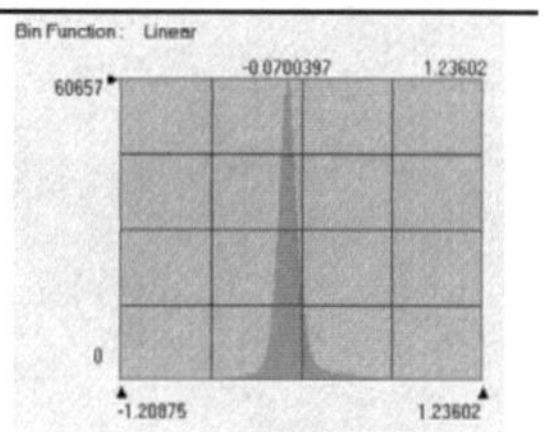
Stat. info	Mean 0.541 Median 0.58585 Mode 0.59862 Std. Dev. 0.179	Mean 0.473 Median 0.50024 Mode 0.49472 Std. Dev. 0.120	Mean -0.070 Median -0.091411 Mode -0.10096 Std. Dev. 0.149

From Table 2, it can be seen that the overall NDVI value is 0.5 which is acceptable for tropical rain forest. In 1996, the mean, median and mode of the NDVI value are 0.541, 0.58585 and 0.59862 respectively. In 2013, all of the mean, median and mode value shows reduction with 0.473, 0.50024 and 0.49472 respectively. The difference of statistic data for both images shows reduction of mean, median and mode with 0.07, 0.091411 and 0.10094 respectively. The mean NDVI value indicates the decreasing of almost 13% of forest health from year 1996 to 2013.

Considering the reduction of forest health due to atmospheric pollutant [2] and air pollutants probed were positively correlated with urban built-up density [3], a set of ambient air quality samples are taken from new road construction at Angsi Reserve Forest from year 2004 to 2008 to see how deforestation and construction affect the ambient air quality thus affecting the forest health. Figure 7 shows the air quality reading during the construction period.

There are 3 stations collecting the air quality data namely A1, A2 and A3 are located at Taman Seri paroi, Ulu Bendol Recriation Park and Taman Seri Bendol respectively. A1 and A2 stations are located near the deforestation area for the road construction project while A3 located further away from the deforestation area.

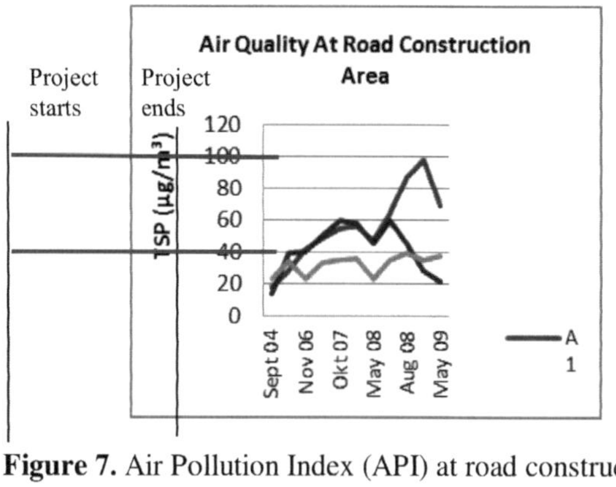

Figure 7. Air Pollution Index (API) at road construction area

Air pollutant index (API) indication: 0-50 as good, 51-100 as moderate, 101-200 as unhealthy and 2001-300 as very unhealthy. The graph shows increasing Total Suspended Particulate (TSP) during the road construction in year 2004 to 2008. The API readings show increasing values to moderate level and there is a time when the value nearly approaches unhealthy level. It proves that air pollution is directly connected with deforestation and urbanization and the reduction of forest health in this study is directly connected with air pollution due to deforestation and urban built up.

4. Conclusion

Based on the remote sensing and image processing analysis, it can be concluded that the forest study area shows a considerable changes from year 1996 to 2013. The NDVI value which is associated with forest health also decreases over the 17 years period. The health of forest is known to be affected by air pollution is tally with API index during deforestation and construction period which shows high API index recorded. The changes of study area are important to monitor forest change so that the deforestation and development can be planned and the balance of ecosystem is still preserved.

5. Recommendation

For future research, the forest change detection can be studied by creating spatial models using ArcGis and its extension Spatial Analyst. Then the factor of changes can be determined and the statistical factor of changes can be analyzed.

6. References

[1] Burgan, R. E.; Hartford R. A. 1993. Monitoring vegetation greenness with satellite data. Gen.

Tech. Rep. INT-297. Ogden, UT: US Department of Agriculture, Forest Service, Intermountain Research Station.

[2] FAO, 2010. Global Forest Ressources Assessment 2010 – Key findings. Rome, Italy.

[3] Qihao Weng, Shihong Urban Air Pollution Patterns, Land Use , and thermal Landscape:

An Examination of Linkage Using GIS Yang Environmental Monitoring and Assessment June 2006, Volume 117, Issue 1-3, pp 463-489

[4] Jitender Kumar 1, Deepak Chaudhary2, Vijendra Rai3. Image Analysis and Processing of Remote Sensing Data with the help of NDVI values International Journal of Engineering Research and Development **e**-ISSN: 2278-067X, **p**-ISSN: 2278-800X, www.ijerd.com Volume 4, Issue 3 (October 2012), PP. 35-39

[5] Sharifah Mastura Syed Abdullah Spatial and Temporal Analysis of Deforestation and Forest Degradation in Selangor: Implication to Carbon Stock Above Ground 2012 4th Conference on Data Mining and Optimization (DMO) 02-04 September 2012, Langkawi, Malaysia 978-1-4673-2718-3/12/$31.00 ©2012 IEEE

[6] Arief Wijaya, Erika Romijn, Saipul Rahman, Martin Herold and Louis Verchot Assessment of Deforestation Drivers and National Carbon Emission Using Remote Sensing Analysis,

[7] Tves Julien, Jose A. Sobrino, Cristian Mattar, Ana B. Ruescas,, Mariam Atitar, Juan Cuenca, Temporal analysis of normalized difference vegetation index (NDVI) and land surface temperature (LST) parameters to detect changes in the Iberian land cover between 1981 and 2001 22085, E-46071 Valencia, Spain International Journal of Remote Sensing Vol. 32, No. 7, 10 April 2011, 2057–2068 978-1-4673-0479-5/12/$31.00 ©2012 IEEE

[8] Innocent NDOH MBUE Landscape Change in the Bamenda Highlands of Northwestern Cameroon: Modeling the Driving Forces of Smallholder Deforestation 2010 2nd Conference on Environmental Science and Information Application Technology

[9] K. C. Tan, H. S. Lim and M. Z. Mat Jafri Carbon Dioxide Distribution over Peninsular Malaysia from Scanning Imaging Absorption Spectrometer for Atmospheric Cartography (SCIAMACHY) International Conference on Computer and Communication Engineering (ICCCE 2012), 3-5 July 2012, Kuala Lumpur, Malaysia

AUTHOR INDEX

Institute of Physics
Dirac House, Temple Back
Bristol BS1 6BE UK

CURRAN ASSOCIATES INC
proceedings.com